Better than Living

by

Dr. Michael Brown

As told to

Scott Locker

ISBN-978-0-9960877-2-8

Cover art:
Universe Woman Silhouette
PickPik art and design
pickpak.com

Author Photo:
Marco A. Martinez de Escobar Martinez
La Paz, Baja California Sur

DEDICATION

To RD, Pax, and Aaron
May your journeys always be courageous.

ACKNOWLEDGMENTS

It is with much gratitude for all those who have helped me to tell this story that I am presenting this history. For those who helped me in the beginning, my parents, my wife, and my co-workers in New York. I hold a special esteem for my colleagues in our clinic in Oregon, their support was the catalyst that pushed us through the doubt and uncertainty of the ethics of what we were doing.

To our first courageous and valiant patients who let us into their most intimate moments at the end of their lives, we would have had nothing to say without you and your bravery in the face of the unknown.

And later, to those of you, especially Max, who hid me, held my secrets and kept me safe, and protected me these past two years, and to Scott Locker, the brave, trusted transcriber of my story, I am especially grateful.

Thank you all, I owe you my life.

Dr. Michael Brown

FORWARD

I met Dr. Brown at a Panaderia (or Bakery) next to a farmer's market in La Paz, Baja California. He introduced himself by a different name then. We found that we shared a mutual sense of preferring the company of locals to the community of North Americans who lived in the area, and ended up forming a friendship of sorts. On our first meeting I translated for the fruit vendor answers to his questions, to which he thanked me, saying he was not accustomed to the La Paz accent, and retired quickly.

I was in the daily habit of writing on my laptop at a café on that block, so twice a week I witnessed the vendors set up their stalls and the arrival of customers coming to buy the organic produce as I toiled with words over coffee and a croissant. Three weeks after my initial exchange with him, he approached me as I wrote and asked if he could share the table. His first question after his introduction was if I was a writer. I told him I had worked as a journalist and now in my retirement was working on my fourth book. We chatted for an hour and then shared a lunch at a local seafood restaurant.

He made a habit of stopping at my table to say hello during his forays to the farmers market every Tuesday and Saturday. After a month of his visits I invited him to come fishing with me. He met me at my home, a few blocks from the café, and we drove my jeep and trailered boat to a launch ramp across from an island I liked to fish at. We launched the boat and cut through the glassy sea to my favorite fishing grounds, a trip of about 45 minutes.

I had always felt that he was a very easy going and confident, yet at the same time a guarded, private person, and this trip started the same way, with him exuding the energy of loving life while holding a secret within his soul. I neither cared nor worried about what was behind this secret, as most everyone carries some form of private burden with them.

He caught two big yellowtail, and I hooked up a Dorado, so we planned a dinner with the spoils of our efforts. After our fine meal of grilled fish, content with our days' work and the satisfying food we had consumed, he looked directly into my eyes and asked with a chilling gravity what I thought about Ed Snowden. His tone had changed so dramatically that I knew there was more to the question than my opinion of the young man who had exposed an NSA domestic spying operation. I paused to process his odd question before carefully replying. I answered that the truth of moral law was more important than the whims of the United States government and its self-serving belief that it was providing security to its citiczens and others by spying on them. I continued to say that the reason Snowden was criminalized was that he was young and on the outside of the "club", and that generals and politicians could do something similar with impunity because they were protected by their masters, and then added that I felt the government's spying was unconstitutional

He asked if I felt Snowden was a hero. I replied that his actions were heroic, his revelation was extremely important, but he was, in my opinion, simply a man of integrity doing what he saw as the right thing by speaking truth to power. He was just the messenger.

My guest quietly took it all in, and eventually asked, with an edge of ire, if I was someone to be trusted. I realized that he needed a confidant, that his secret was a huge weight for him to carry alone. I told him that, and added that I didn't care what he had done or who was after him, and that if he had betrayed national security or robbed a bank, that was his issue, and since he had not harmed me as far as I could tell, that I had no reason to betray his trust. As a journalist, I continued, I knew the value of protecting a source and a secret. He remained quiet for a long time, then excused himself, brought his dishes to my kitchen, and thanked me for a wonderful day.

Two weeks later he found me once again at my usual spot and invited me to invite him fishing again. Though I sensed fishing wasn't his priority, I acceded and made a date for the following morning.

We took the boat out, and once the lines were in the water he got to his point. He asked if I would be up for writing a book for him. He said he had a story to tell that needed to be told right away. That it was crucially important and that he had an immense amount of documentation to back up his tale. I asked him if his story had anything to do with national security or the Snowden affair, and he chuckled, replying that it was in many ways similar, but no government was trying to throw him in prison; they were in reality trying to *keep him dead.*

So now that he had my attention, I suggested he start talking and let me decide if I wanted to involve myself. He asked me to swear secrecy, to which I agreed if I refused to write the book. If I decided to write it, however, I'd tell nothing but the truth, and I'd need to confirm everything he would tell me.

He replied that it was pretty much all public record, except for the past two years, which weren't germane to the heart of the story. He looked once again into my eyes and extended his hand, saying that if we had a deal, my world would very shortly be rocked. We shook hands on it and he fished through the cooler, pulling out a couple of beers. After a couple of sips he said, "My name is Doctor Michael Brown, I scientifically discovered and proved what happens to our soul after we die. I imagine you may have heard about my research. You may also have heard that two years ago I was killed in London. That report was somewhat exaggerated, as you can see…"

Obviously, I was amazed and entranced with the man. After he finished his abbreviated version of his tale and I regained my composure, I laid out the rules of my writing his story, which we stuck to. He reached into his backpack and handed me a folder with sixteen SD cards inside of it. He said they contained a draft of his story and extensive notes on many of the details of his work, as well as thousands of pages of scientific data, audio, and video recordings.

We agreed that he would tell me the story first hand, verbally, so I could turn his meter and tone into the emotion and passion that this stack of notes and scientific documents would likely not reveal. I wanted to breathe life into it all. I recorded all his interviews and transcribed them faithfully, editing only for grammar, clarity, and flow. The following story is 100% his. I added nothing, except where noted. We met almost every day for six weeks. His words flowed easily and I asked few questions of him. Every night I went to my computer to check the historical record of what he had told me (in the encrypted way he asked me to, so as not to draw attention to our whereabouts), and always found collaboration with what he was saying.

Once I had transcribed everything, he read it and gave me his blessing to publish. He then moved on from La Paz, giving no clue as to where he would go nor end up. I've since heard from him twice; once to check on progress and again to confirm his identity to my publisher via skype, and since then, nothing. The following story is told directly from Dr. Michael Brown, and is all true.

Scott Locker

La Paz, Baja California

Better than Living

By

Dr. Michael Brown

As told to

Scott Locker

1.

I am in hiding, writing this as a testimony to the truth of my discoveries and their repercussions, which have been following me these past seven years. Unless you live on a remote island or in the depths of the Amazon basin, you have surely heard of my remarkable findings and the knowledge and devices that gave people choice and control over their lives. The story started out being reported somewhat accurately, but later got distorted and twisted until lies permeated everything I attempted to bring to the public. There has been a great deal promulgated trying to discredit me, and an equal amount it seems, trying to keep the truth pure. But unfortunately, the followers who are defending my discovery and its message are also adding their political, personal, or religious biases into their versions of the truth.

I am here writing to set the record straight, as it were, once and for all, not only to rebut those trying heartily to destroy me and my discovery, but also to leave the truth uninfluenced by the misguided assumptions of those who have been corrupted by the lies. I am safe where I am, and am writing this document of my own volition, although I was asked by my most ardent supporters to do so. My publisher has agreed to publish the result of these efforts without editing, and simultaneously in every form of free and open sharing media in the world to insure what I have to say does not once again get corrupted or buried.

Please note that I don't want to be anything but left alone. I am neither a hero nor a martyr, cult leader, religious icon, guru, nor anything else our culture or society might want to label me with. I wish no fame, nor to be revered or demonized for anything. I am only a humble messenger, nothing more. When you read this document, please realize that any quotes are usually taken directly from our (my team and I) detailed notes and the thousands of pages of laboratory documents we compiled. Any quoted conversations outside of the clinical setting come from my exceptional memory and are as accurate as humanly possible without dependence on a recording device. The point is that I am writing the truth down as objectively as I can, and given that I am a scientist, I promise its accuracy.

The chemical formulas I used will be presented in an abbreviated form in this text so as not to burden the casual reader with the minutia of the process and the often-monotonous pharmacology of that subject matter. I will include all technical material as a free and open source to everyone in the endnotes of this narrative. As there are several hundred pages of technical data, they will be listed as volume 2 of this document, but I assure you, as does my patron, that they will be readily available to all.

I have included herein the events that in my eyes were key and crucial catalysts in putting me on this path. There are details, which may seem to some readers somewhat irrelevant, but to me they helped form the foundation of who I am now and how I got here. In other words, I have included the pertinent incidents in my life that motivated me to do what I did in the name of helping mankind.

In this document I will present all names as they were told to me. If I feel the need to protect someone, I will use only their first or middle name to address them. The details may at times seem unbelievable, yet I assure you they are all completely true and accurate. I write this to keep the freedom of mankind alive and to give hope to those who fear the unknown. There really is nothing to fear. If we are finally able to remove this fear from society, it will liberate us all. I see a future without wars, a future without the elite dictating how we live or end our lives ever again, a future which puts us all on a level playing field in many ways, because a life without fear is a life of infinite possibilities.

2.

My name is Michael Brown. I was born in Los Angeles, California in 1967 and was fortunate enough to have been blessed with parents who had the foresight and resources to insist on my having the best education possible. Through my formative years I went to private boarding schools and was taught many subjects one normally doesn't have the opportunity to study in lesser situations. By the time I had completed high school I had mastered Latin, French, and Spanish, as well as advanced chemistry, physics, and mathematics. I scored among the top three percentile of the country the first time I took my SAT and ACT college entry exams, which I took as a junior, and was offered scholarships to more than a dozen universities that Summer. I still had a year to go in high school though, and still hadn't found my true interest for the direction of my life. I was fifteen years old that Summer, and after conferring with my parents, decided to complete high school and take a few months off to see the world before accepting one of those scholarships. We felt that the offers would keep coming in if my grades continued as they were, and the break would help expand my horizons and assist me in choosing a career.

After graduation, my father had given me an itinerary which included all the highlights of Europe, along with a plane ticket and a Euro rail pass. I explored the old world like a tourist, and after a while felt the desire to do some serious and different exploring. I abandoned my travels in Europe and was able to find passage on a cargo ship bound for South America, figuring to see Brazil and Argentina before returning home.

While on board, I fell in love with the world of ships and the slow yet steady pace of the cargo trade. I learned that the ultimate port on this voyage was actually Iquitos, Peru, more than 2,300 miles up the Amazon River, and asked if I could stay on until then. I was warned by the captain that the interior wasn't the most hospitable environment for a young white man who had never felt callouses on his palms, but I was not to be deterred.

Our first port of call in Brazil was Villa do Conde, in the bay of Marajo, about 500 miles inland. While watching the offloading operations form what I thought was a safe perch over the cargo hold, the tarp I was sitting on was pulled out from under me when it snagged on a cable from the cargo crane, and I tumbled headlong into the hold, about 30 feet below. In the subsequent impact my body had suffered a traumatic shock, which temporarily disconnected my brain from everything below in an effort to protect itself from further injury. I was, in that brief period of shock and tremendous trauma, dead. My world went dark as my body was turned off. I felt nothing.

Then, a brief moment later, a warm glow came over 'me' and I was enveloped in warmth and love. The warmth developed into a "visible" glow of warm orangish-red, and I heard a comforting humming or indistinct voice singing a song I had never heard before. It was instantly warm and comfortable, like an infant being cradled by her mother in the dark. As soon as the blackout hit I was at peace and happy; the mounting pain had vanished in a flash. There was no sight as I knew it, but I could see in another way, like a sense. I felt no body, but still felt whole, the warmth spreading outward quickly. I felt powerful feelings of Love and empathy entering into what I was becoming. It felt as if I were on a launching pad, of sorts, ready to spring off onto a journey into what felt like the cosmos. I was poised to go, without reservation or fear of any kind, anticipating something marvelous, when I felt an insistent tugging from where I had been a few moments before. I was somehow being pulled back, slowly returning to the physical plane, and started hearing the voices of the ship's crew.

The darkness/light started lifting, and with it the warmth and comfort, and my earthly senses slowly returned. The white-blackness faded to red, then the faces grew clearer, and I could hear the people over me calling my name and trying to elicit a response. I had felt a powerful sense of joy and peace where I was and a powerful desire to stay there. I tried as much as I could to hold on to the previous place, but it wasn't working. It all-too-quickly faded as I was returned to the default world.

Now I could feel and focus and speak. The others around me, the ship's crew members, helped me up and over to a place where I could be examined, where I felt the last vestiges of the other world fade out of my memory as if it were but a dream. I was back from the other side. It felt like I was there for a while, maybe 20 minutes, with no real sense of time per se, but my attendants assured me I had only been gone for a minute or two. As I regained my bodily control, I felt a sense of longing to return to the place I had been, and that feeling stayed with me for several weeks, until the vividness of the memory finally faded to shades of grey. I didn't feel my body hit the deck like a felled tree; I didn't hear the people rush to me and try to comfort me, I was totally gone from my body in those few minutes, to a place which felt more loving and home-like than anywhere else I've been.

Amazingly, there was no permanent damage from my fall. I healed quickly, and after more than three weeks aboard the ship, seeing about the same amount of river as ocean, I disembarked in the Peruvian jungle to see a primitive and little-known world. My travels took me through Peru, Ecuador, and Brazil, mostly by one form of watercraft or another.

One evening the river boat I was traveling on was flagged down by an American and two native guides standing on the shore. The captain pushed the shallow bottomed craft up onto the shore to pick up his new fares, and that night I met Loren Miller, a self-proclaimed medicinal plant scientist who, over a few beers as the steamer chugged through the night, told me of some of his amazing research and discoveries in the Amazonian rain forests.

After he spun his background and how he came to be hailing a riverboat in the middle of the jungle, he reached into his backpack and pulled out with one hand a knotted and twisted chunk of a vine, about a foot long and two inches in diameter. With his other hand he held a fistful of bright green tropical leaves and twigs. He said he'd heard of the vine and its magical and medicinal properties for years in his research, then finally had drank the extract under the guidance of a local Shaman, and between the research and his first-hand spiritual experiences, felt that the benefit to the world would be exceptional once all the research was completed back in the states.

Miller told me that the local people had used the plant for around 6,000 years as a medicine and an instrument for the Shamen to guide their tribes through life and death. He said there were thousands of accounts of spirit guides awaiting the person entering the psychedelic voyage to whisk them away on a journey through the stars, and that the plant answered even the most profound questions posed by its guests. And in most instances the voyager comes out of the trip with a sense of illumination, elation, and sometimes a spiritual transcendence.

The plant contains a psychedelic compound, which creates spiritual visions and astral voyages in the users, but to me the most amazing part of the story was that the extract of the plant, if ingested orally on its own, has absolutely no effect on the user. Somehow way back in time the indigenous people found that mixing the vine with another plant, also on its own completely inert, would protect the important active ingredient from getting destroyed by the user's digestive system, thus yielding a pure and powerful psychedelic or mystical experience. Loren told me that the original discoverer of this ancient combination, an elder Shaman who lived thousands of years ago, claimed that he was told in a dream to mix the two plants together in specific proportions and to drink the tea to find the pathway to the soul. Another version of the story recounts that the plants themselves came to the Shaman and told him how to marry them into a brew.

Miller explained that the vine, Banisteriopsis Caapi, was a natural monoamine oxidase inhibitor (MAOI), and the other plants, such as Chacropange or Chacruna, contained high concentrations of Dimethyltriptamine, or DMT. He told me that DMT is a naturally occurring compound in thousands of plants, yet as we consume them our digestive system oxidizes the DMT before it can be absorbed into the blood stream. With the MAOI ingested with the DMT, the digestion is neutralized and the DMT can pass unmetabolized through the membranes of the stomach and small intestine, and eventually cross the blood-brain barrier. The science of it all was very technical, yet the Indigenous tribes of the upper amazon basin have had it figured out for millennia.

As Miller enthusiastically told his story and exhibited his treasures, I saw him as a modern-day Indiana Jones, and at that moment my life's course had been set. I drifted in and out of his adventures, blending them with visions of my possible future of finding cures for illness through plants like these, refining chemicals from nature, and discovering something that would aid mankind away from its suffering. He told me there were thousands of undiscovered species of plants in the region we were passing through at that moment, not to mention the potential in classifying those plants and testing them for unique chemical compounds that may benefit mankind. My head soared imagining the possibilities around me as I fantasized about a future in the jungle and research in a fancy laboratory somewhere nearby.

Two days later he left the steamer in search of other possible treasures. I was left alone with my imagination, and soon became anxious to return home and start on what I saw as the new course of my life.

While on our way down river at a stop in a small village, one of the crewmembers, while attempting to moor the boat to the village dock, slipped and fell into the water, and within a few seconds let out a piercing scream. The crew went to retrieve their companion as the passengers stared on wondering what he had encountered under the water. As he was hauled onto the deck, everyone watched in shock as huge welts and blisters began to form and rapidly grow on the man's legs, arms, and torso. The man began to convulse as he pleaded for help between screams. Someone pointed out a patch of bright colors where he had fallen, and then I saw they were not flowers but extraordinarily colored little frogs. I heard the captain explain to the crew and passengers in Spanish that his poor crew-member had fallen onto a colony of poison dart frogs, followed by the somber statement, “He’s going to die, without a doubt he’s going to die”.

Some valiant soul wrapped him in a blanket, and as much as we tried to turn away, the suffering of that man held us all entranced. He lasted more than a day before finally expiring. I was wracked with a sense of futility at having borne witness the painful demise of that crewmember. It was my first taste of death and it left a profound mark on my heart. I wanted in some way to comfort the man, but the captain had placed two guards at his side to insure nobody would touch the victim, lest the poison be transferred to someone else. The next afternoon the boat nudged onto a large sandbar and three of the crew hopped off with shovels and dug an impromptu grave for their mate. His canvas-wrapped body was quickly laid to rest without much in the way of ceremony and within minutes we were again on our way down river.

I thought of the fascination of the jungle, now tempered with deadly danger, but also of the possibilities of harnessing a poison like that of the poison dart frog might bring to medicine. The stakes of my future trek through the jungles were significantly raised when I witnessed the death from such brief contact with those little creatures. I also felt a huge dose of empathy for the man and his suffering. If they knew he was going to die, why couldn't anyone have mercifully put him out of his misery? Wasn't there a compassionate or morally justifiable way to swiftly end the suffering of those cursed with an agonizing death?

Three weeks later I was on a plane from Rio De Janero to Los Angeles, looking forward to telling my parents of my decision to make a career of medicine and medicinal plants. As I researched the fields associated with discovering medicinal uses for plants, I found many paths available in an assortment of disciplines. My parents suggested I start with an education with a strong medical foundation and go from there, which to me made good sense, and after three years of undergraduate studies at Stanford, I was accepted at the Columbia University school of medicine in New York.

As all too often happens in life, my dream gradually shifted as other interests entered my life. There were two, to be exact, which diverted me from returning to the Amazon with a pith helmet, butterfly net, and a pair of pruning shears, the first was a fascination with anesthesiology, and the second was Olivia.

3.

Somewhere in my third year of medical school we were taught about the God-like control the anesthesiologist has over the patient as he adjusts the flow of vapor or IV drip into the blood stream. I became entranced as his every adjustment of the dial was reflected in the vital signs of the sleeping patient. Anesthesia comes from the Greek root, and means, “without sensation”. Although the anesthesiologist usually does not directly provide a benefit to the patient (except in cases of pain management), he directly helps those who are helping the patient, such as the surgeons. In most surgeries, the anesthesiologist bears the grand majority of the responsibility for the well-being of the patient, even though his contact with her is generally minimal. I immediately wanted to learn all I could about bringing patients to this near-death state where they felt nothing, remembered nothing, and were brought back to a conscious state seemingly with a wave of the anesthesiologist’s hand.

And then I met Olivia. We were standing next to each other on a New York subway as it headed south towards midtown one Friday night. I was supposed to meet friends for dinner and drinks; she was looking tired like she had just finished a thirty-hour shift in an emergency room. A seat was vacated behind her and I asked her if she would like to sit down. Olivia looked at me and said, "Aren't you the chivalrous one"? I took that as an opening to introduce myself, and we started talking. It was effortless and captivating. After a few minutes, she asked where I was getting off, and I replied, "34th street".

"You know that was like four stops ago"? She smiled as if I had been absent minded.

"Yes, I do", I answered. "And to be honest, it doesn't matter. I can't imagine anywhere else in the world I'd like to be than where I am right now at this exact moment". Her smile faded, then returned with a completely different message. She said, "I'm Olivia", and extended her hand. As we talked we found out that we were both about to start the residency program at Columbia in a few weeks. I couldn't believe I hadn't noticed her before. We were both in the same program in the same semester at the same school. Olivia wanted to be a surgeon. We compared our resident courses and schedule and found that we would be bumping into each other frequently in the coming semester, so we made a pact to become study partners right there.

Olivia and I had a wonderful four years of residency, mostly together, and married a few months before we were to be set free from Columbia University's Saint Luke's Hospital residency program into the world of being full-fledged doctors. I was accepted into the Anesthesiology department at Mount Sinai Hospital on the east side of Manhattan, and she stayed on at Saint Luke's, which is located on the West side and 118th street, and we started nesting in an old building just off of 100th street and Columbus avenue.

Despite the intensive training and the obnoxiously long hours of our young careers, we found a love more profound and entwined than anything I had ever seen or experienced in my life. Our moments together were precious and cherished. We never fought; we just loved more and more. Our friends always asked us how we did it, the *it* being our ever-expanding love for each other despite the outrageous pressures of our careers. We had a few answers for them, but the reality was that we were just an exceptional fit, we communicated well, and we respected each other without judgment. Our bond was intractable.

For our fifth wedding anniversary we took a couple of weeks off to explore the peninsula of Baja California, flying in to San Diego and renting a car to make the 2,400-mile voyage down this long desert and sea peninsula and back. It was such an amazing adventure that two years later we decided to return and concentrate on the places we had enjoyed the most. We were swimming in the Sea of Cortez and eating fresh caught lobster while appreciating all the good fortune we had been given in our lives. We were now out of debt from the student loans, and in our celebration of life on this vacation we decided that we were ready to have a child or two. Olivia said she wanted to get started right away.

The next night I was playing with her in bed, doing my best to tickle her while she did her best not to giggle, when I noticed a pimple on the instep of the bottom of her right foot. Although pimples are an ordinary occurrence, this one held my attention because of its location and color. The skin on the bottom of the foot is not built to be conducive to acne, and the thing itself was dark without redness. Olivia said she felt nothing, so we decided to leave it alone and address it once we returned to New York. I monitored it daily and noticed that it was growing. We had hoped it would have started to recede but that wasn't the case.

Back in New York we had a Doctor friend look at Olivia's foot. The pimple had grown in a week from 3mm to 9mm, so the Doctor cut into it and drew the dark pus for testing. The wound was much deeper than it was wide, and he ended up packing it with a bit of cotton before bandaging it. We waited with anxiety for two days as the tests were being processed. Although the wound wasn't infected and didn't hurt, its oddity gave us cause for concern.

One of the benefits of being a medical professional is that our doctors keep nothing from us, and when the results returned, our friend called Olivia immediately and gave her the news. He said it was a cancerous growth, and that we needed to see an oncologist when we could. Another privilege of our job was that we knew the inside scoop of who was the best in their field, and being a doctor herself, Olivia got preferential treatment. There was none of the usual waiting six weeks for an appointment; we were seen the next day by one of the most qualified specialists in the world, where he had her go through a battery of tests and imaging procedures. After his analysis, he called us into his office and gave us the news.

Olivia had developed T-cell prolymphocytic leukemia, and her probability of survival to one year was less than 20%. It had silently metastasized to her organs and muscles. He recommended she start chemotherapy as soon as possible. He also suggested she clear her surgery schedule for the course of treatment, as the side effects would be an impediment to her specialized type of work. She was stunned at the thought of not working and the projection of what her future might bring. She was 36 years old and approaching the apogee of her life, an excellent surgeon living a story-book existence with the love of her life, possibly even pregnant, and now in a matter of days it had all gone awry. I tried to keep her focused on the positive parts of her life, but we were too well trained to ignore the reality of her disease. Being in excellent health and a fighter, she was determined to beat the odds.

Treatment started and she reacted poorly to the chemo. She grew pale and lost her hair within a couple of weeks. Although she was accustomed to running in the park and finding time to exercise whenever possible, the reserves of strength she had accumulated were soon depleted. In three months she had lost one third of her body weight, down from 155 pounds to 102, as the cancer ate away at her insides relentlessly. At six months we were faced with the reality that the metastasis of the disease was barely slowing despite all the measures we were taking. She went through surgeries and three rounds of chemotherapy, and was taking 62 pills every day. Her pain level was relentlessly climbing and now she was hurting all over, all the time.

I was able to prescribe appropriate meds to help her

sleep through the pain and anguish of her nighttime suffering, but that was only a temporary pause in what we both realized was now her journey toward death.

A few weeks later Olivia whispered to me that she wanted to die. She said she knew there was no hope, and she was tired of suffering and rotting away while awaiting her final breath. I discussed it with her as best I could, but she had thrown in the towel. My not wanting her to go was, I realized later, my own ego demanding its way with me. My selfish ego, and the law of the state of New York both dictated my course of doing nothing except watch her suffer and suffer and suffer more each and every day. She was passing dark and putrid blood from her bowels and occasionally vomiting it up as well.

A week later, on the morning of her 37th birthday, she had regained the semblance of a smile and a bit of energy. As I brought a cup of juice to her lips she shook her head and asked for a kiss instead of her liquid breakfast. I bent down to her bed and we put our arms around each other. As I kissed her cheek she tightened her hug and whispered: "Please kill me dear. You can do it so easily and nobody will know… I'm already dead and we both know it. Please don't force me to suffer any more. You can easily give me a few puffs of Propofol on the vaporizer and I'll be able to go to sleep for good. Please, please do this for me". She let her arms slip back to the bed as her strength waned and started coughing up more bits of blood until she passed out a few moments later.

I was appalled and horrified at her request. As much as I wanted to give her peace, my life had been dedicated to saving and prolonging life at all costs. Now my love had asked me to stop her heart and release her spirit. I didn't know how to do that, how to give up on the woman I cared for more than everything else in the world. The legal aspect came to mind later and I trembled at the thought of being convicted of killing her and spending years in prison for manslaughter. Of course, my rational mind would have rebutted that reasoning with the facts that there would likely not be an autopsy, and that if there were to be one, the residue of the anesthetic would have been almost impossible to detect as it quickly evaporated from the lining of the lungs and didn't linger in the bloodstream except in the few cells that were stuck with the compound on their way to the brain after her heart would have stopped. The odds of being caught, prosecuted, and convicted were almost nil, but my ego, in the guise of a moral compass, was the heavyweight in this battle within me, and it told me I could not let go of my darling wife no matter what.

Three hours later she woke up. Olivia looked over at me as her face turned into a twisted scowl and softly exclaimed: "I'm *still here*. What's wrong with you, you selfish pig fucking coward? What did I ever do to you to make you torture me like this? I want to go and you just sit there searching for the balls you'll never find. Get out of here and don't come back, you bastard! I want a divorce!" I had never heard her use profanity. My soul was crushed, and as I left the room I felt ashamed of my weakness and imbedded sense of sanctimonious righteousness. Her voice, weak as it was, screamed to my soul powerful sentiments perfectly crafted to pierce my every defense and rationalization.

With each successive breath she took after that I was forced to reevaluate my way of thinking, as well as the arbitrary laws levied on a person's rights during their final moments on earth. I had been told throughout my existence that life itself was paramount and needed to be preserved no matter what, that taking the life of another being was wrong, that suicide or killing was a sin, and that the laws of the State were there to protect and insure the well-being of her citizens. Now as I contemplated my beloved suffering beyond all reasonable means of acceptability these teachings and laws seemed to morph into some kind of big lie; the kind that gets repeated so often and for so long that society finally accepts it as the truth.

I went back to her room to apologize and try to

comfort her, but as I opened the door I was greeted with a wag of her finger and then a dismissive pointing of it towards the door. I retreated and started to weep. This was our first fight. I returned later and thankfully she was asleep. I held her hand and suspected that if she had been even a little bit more aware, she would have pulled it away from me. As was often the case at that time, I rested my head on the bed as I sat in the chair next to it, and fell asleep. I awoke to having it pushed until I almost fell over, and as I regained my balance I saw her hand balling into a fist preparing to whack me. I retreated once again.

50 hours later she passed away. She was whispering screams of agony through most of her last two days, yet she still had enough control of herself left to die with her middle finger extended, to insure I got the point that she was still pissed off at me. I held her through her last breaths and heartbeats, and then she was gone. Her body was now a bag of tissue and bones; her mind was now just a brain. She had become a cadaver in my arms. I felt the life flow from her body as she passed on to who-knows-where. What had a moment earlier been my wife was now something completely different that had neither animation nor spark.

4.

Riding the grief of losing the love of my life, I plummeted into a cavernous depression where I remained lost and searching for reason and understanding. I asked myself why she died, why she had to suffer and why I was now alone. I couldn't even find the guts to clean out the cancer medicines from our, now my, bathroom and bedside table. The apartment was left as it was the day she died for the most part. I often closed myself inside of her closet trying to recapture the essence of my wife in the scent of her hanging clothes and the sight of what she used to wear. After three months of numbly stumbling through work and wrestling with the desire to check out of this life myself, I found no reason in her passing, her leaving me alone, and most of all her having to have suffered so horribly before she died. I struggled without success to rise out of the depths.

At six months I was almost no better off, and work had completely lost its charm. I alienated myself from my friends and hardly spoke with my family anymore.

Olivia's final words constantly replayed themselves within my mind and I tortured myself relentlessly for my part in her suffering. Eventually my parents came to New York to try to pull me out of my self-imposed castigation. While mom went to work reorganizing the apartment and editing out Olivia's stuff for donation, my father started discussing options with me. After a couple of days, I was able to let their words and actions penetrate. Olivia's possessions were boxed up and discreetly removed, and the possibility of somehow rebooting my life started to weakly impregnate my soul.

A week into their visit the funk within me started to crack. I was returning home from a run and had mindlessly paused in front of a neighborhood produce market. As I stared blankly at the brightly colored vegetables, my mind seemed to turn itself off, as if I were meditating or in a trance. I must have stood there for five minutes, mesmerized by the vegetables without a thought in my head until I was jolted from within as if I had been touched by a cattle-prod. I gasped and reanimated myself, and as I once again focused on the display of polished produce, I became inspired to make a dinner for my parents.

As I filled a basket I became excited about cooking that meal. It had been more than a year since I had felt any enthusiasm for cooking, much less anything else.

Now it felt like the cavernous void I had created was being filled with ferocity. I saw endive, watercress, and arugula and Meyer lemons marrying with each other as a bright and crisp salad (the lemons as part of the dressing), and then grabbed a bunch of root vegetables to roast as an entrée, followed by baskets of fresh berries to be baked up in a galette for the final course. I found a bottle of dry Riesling to accompany the meal and discovered that there was an emerging energy in my steps as I carried the shopping bags to the apartment.

While we ate, I started telling my parents of my new plans. These plans were so new that they were erupting from my mouth without any forethought, coming fourth as true spontaneity. I thanked my parents for their concern and council, and told them that they were correct in suggesting I get away for a sabbatical. I retold the story of my Amazon adventure and how the jungle had turned the course of my life. I wondered where I would have ended up if I hadn't become an anesthesiologist and met my wife. The course I had previously chosen now beckoned my attention once more. I had decided (at that moment) that it was best if I took some time away from New York and went on a trip of rediscovery.

Maybe it was a release of years of subconscious repression, or perhaps some higher power taking control of me when I was too depressed to think, but in any case, a month later I had sublet my apartment and taken a leave of absence from my duties at Mount Sinai, and found myself back in the Amazon basin.

I had volunteered to assist a Brazilian group looking at the rain forest for possible clues toward cures for cancers and other diseases. The mission was headed by Dr. Drauzio Varella, an oncologist who had lost his brother to lung cancer. The research was sponsored by San Paulo's Paulista University. We worked for six months collecting and cataloguing samples of all manner of flora on our floating laboratory, an 80' river boat converted to this special use. We were on a quest for medicines of the future. It was a period which brought me to a point of focus regarding the future of my life and career.

Our search was often suggested by observing the local tribes and their use of plants as medicine. Dr. Varella told me that when a plant compound had shown benefit, its rights were immediately bought up by one of the many pharmaceutical companies waiting at the doors of the university. And although these purchases maintained the research, the goal of these companies was in contrast to his beliefs that the medicines from plants should have no proprietary ownership. When a drug company synthesizes a plant compound, he said, they steal God's gift and replace it with profit.

I decided to investigate the Ayhuasca I was told about (when I was there as a teen) by Loren Miller. I took an unauthorized side trip to visit a Shaman and participated in the ritual. The experience was truly fascinating in that it had been used as a sacred ritual for millennia. I observed the Shaman as he pounded the bark and brewed his spiritual tea, all the while chanting and singing. I had been fasting and meditating before drinking the brew, as instructed. We were in a jungle village of a few huts and a muddy clearing which delineated the pueblo. Wildlife surrounded us and announced its presence with a cacophony of chatter and song. As the sun set we drank the concoction and listened to the song of the birds and the Shaman as he gave rhythm to the concert above.

When the DMT crossed my blood brain barrier, I felt myself being lifted into another reality. The analytical me soon dissolved as I was completely taken up by the sensation and imagery that consumed my being. There were no classifications or boxes to fit my experiences into, so I let go and entered into the journey as a willing participant, leaving the clinical observer behind.

This journey came on as a feeling of warmth and love. I was not alone, but surrounded with what I would call ‘guides’. At first there was nothing to see, just a sense of dark warmth and gentle encouragement to let go and join in. I felt elevated and then it was as if I were flying through space. Soon the visuals started. I was being escorted by jungle cats and tropical birds, running and flying beside me as we raced through, and then above the jungle. We soared into the night sky and into the cosmos. I experienced connection with colonies of like energy all through the stars, all welcoming and loving, supportive and compassionate. I seemed to fly for a couple of hours, fearless and smiling, connecting always with these small 'beings', which felt like shimmering yet invisible balls of energy, about 2-3' tall. They were all happy to 'see' me. I felt empowered.

Then I 'saw' Olivia. I felt her energy envelop me and hug my soul as we flew together. She spoke to me and told me that our connection and love transcended the limitations of Earth and being alive. She told me that she forgave me for my selfishness in not letting her go when she needed to be rid of her body. I was overwhelmed with sensations of joy and love. There was no remorse, no guilt, no sadness, just love. Later in the voyage I came to a bright band of stars which I took to be an angel or God, a being who was all-knowing. I asked a question: "why am I here, I mean, what is my purpose"?

Her reply: "To love, silly!", and then she retreated into the cosmos. Soon the stars started to fade, the jungle animal guides returned to bring me back to Earth, and the journey had peacefully ended.

I felt I had been enlightened. I had seen what lies beyond our mere mortality, our eternal universal connection. The experience left me pondering our post-living state. When I discussed the trip with the other participants the following morning, many reported similar experiences; the animal guides, the feelings of being surrounded by others (some described the beings as family members), and a pervasive feeling of being loved. Rationally I knew that the drug had interacted with my 5hd1 serotonin receptors, but the emotional rest of me knew that this was a message (and an experience) of great significance.

5.

After my sabbatical in the jungle, my work as an anesthesiologist resumed at Mount Sinai in a slow and disconnected manner. I felt as if what I was doing had no promise anymore. I grew bored and restless and eventually realized that I needed to be doing something else. The sum of my experiences with Pain and death shouted to me of a different calling, presently unclear yet haunting in its insistence.

My uncle, Robert Brown, had invited me to come to Oregon and live with him on his ranch in the Hood River Valley, a few miles outside of Portland. He had 500 acres and a couple of out buildings I was welcome to use and enjoy for as long as I needed. He had described the place in his mild-mannered way as the cure for what ailed me. I knew that the depression and lack of motivation I was inundated with needed to be shed by whatever means possible, so I decided to give it a try. The Amazon hadn't cured me, nor had returning to work, but maybe being with family in a new and dramatically different setting would yank me back into my life again.

It was time for a change, so I sold what I could, gave the rest away, and with two suitcases I boarded a plane bound for what I hoped would turn into a new life. Oregon was a beautiful counterpoint to the cold concrete darkness of New York. The ranch was green, bathed in sunshine and open, and I could walk barefoot through the grass while breathing pure air whenever I wished. Fruit orchids filled most of the property, so my diet improved significantly as well. As I settled in I felt another huge block of my sadness fade. The area was beautiful, the people friendly, and the energy there was completely different from my previous residence. I smiled for the first time in a long while.

When uncle Bob saw my emotional improvement, my being so at ease in this new environment, he suggested I permanently relocate to Oregon. He said I could either live on the ranch with him or move to Portland, but he quickly added that since it was only 50 minutes away and the rent on the ranch was so cheap (meaning my chores), staying was the best option. There were two other houses on the property, either of which I could choose to use as my residence or an office for my practice, or one for each. And besides, we could keep each other company.

He had been a harbor pilot for the commercial shipping transiting the waterways around the port of

Portland and had retired two years prior to my arrival. He had been married for 30 years when his wife abruptly announced one bright summer day that she was leaving him. Bob was completely taken by surprise at her decision and watched numbly as she got in her car and drove off the property and out of his life. He had been alone since, still trying to figure out what went wrong. She didn't go into details in their brief discussion that afternoon, so he thought it might have been a temporary thing, but seven years later he was still scratching his head and asking himself why she took off.

We spent the summer evenings that year sitting on his front porch sipping lemonade and commiserating on our losses, our triumphs, and just about everything else in between. I felt the pain of my loss morphing into purpose and slowly came to the realization that I could do good here by working with the sick and dying and help them bring dignity to their last days My uncle reminded me that assisted death was legal in Oregon, so my research or practice or whatever would probably not be impeded by the law. He became my sounding board as we discussed the issue and how do make a difference with folks who were tired of living.

I refurbished one of the houses Bob had offered me and, once I had gotten my license to practice medicine in Oregon, hung out my shingle and started my new career as a country doctor. The development of my practice was slow yet steady, which was fine with me. I set up a general practice on the ranch and between patients dove into the research I had started earlier in the Jungle. Of all the plants I studied from the rainforest, none had as much value as the Ayhuasca that I had been exposed to as a teenager. There were now lots of big companies scouring the jungle for the next new cure for something, and following the progress for me was fascinating.

I adapted to the small-town way of life and found my inner rhythm. I was finally happy again and pretty much as recovered as I could be from the loss of Olivia. I had no desire to replace her, so stayed home or went out with my uncle, and that was fine. Three years had passed since my move from the big apple. It seemed that I had found peace at last.

Bob would visit my little home office regularly to bring us food and shoot the breeze with my staff. He usually showed up about once a week and had lunch with us. Sometimes I would be busy with a patient, but it didn't matter to him. There were four of us working there at that time, Sally, the receptionist and office manager, Jenny, my nurse, and Eric, my research assistant. Bob would chat with them and leave just before the office would reopen after lunch, but this time he stayed and asked if I could give him a check-up.

It's funny to think that in three years of our living with each other, I never insisted he come in for regular check-ups, but because of his strength and the quiet manner in which he went about the endless chores on the property, I never did. I guess I just took it for granted that he must have had his own doctor. I admired him and the natural fitness program the repetitive ranch work had instilled in him. I watched in mild amazement as he chopped firewood and tended to his crops and livestock. I never thought to caution him on his petty vice of an occasional cigarette and brandy, which he indulged in maybe twice a week. And although he was twenty years older than me, I never looked at him as old. He was simply my ageless uncle Bob.

"I think I might have come down with the flu or bronchitis or something", he said. "I've been coughing and wheezing a little these past few days, feeling tired and weak with shortness of breath and achy bones and joints". I replied that it did sound consistent with a respiratory infection, and since he was finally in my office I was going to give him a full checkup nonetheless, and assured him that we would have him feeling better in no time. I asked Jenny to draw blood for a full workup and asked Bob to let me listen to his lungs. Through the stethoscope I heard abnormal sounds of congestion and wheezing, and told Bob that we needed to get an x-ray of his chest to better determine his problem.

The next afternoon he came into my office with his x-ray in hand, having gone to a lab in Portland for the imaging. As I pushed the film onto the light box hanging on my office wall I saw something I had not expected. There, staring back at me from the image burned onto the film were eight distinct tumors. One occupied the left lung by itself and was the size of a baseball, and the others were dotted throughout the other. I stared in disbelief at the spots on the film, waiting for them to disappear as the light adjusted. My smile vanished as I came to the realization of the probable diagnosis. I was frozen in place, finally coming to my senses when my hearing registered my uncle trying to get my attention.

"Mike, is something wrong? You all right there?"

"Yeah uncle Bob. From your x-ray, it looks as if you have more than just bronchitis. I'm going to schedule a CT scan for you right away, and then, if needed, a biopsy, OK?" He looked confused, so I took a second to regain a bit more of my composure and explained: "What this light spot, rather all these spots indicate are the possibility of tumors. The CT scan will give us a clearer picture of them and from there we can make a better determination as to if they are benign or not. If there is a question of their being malignant we'll have a biopsy performed to determine what they might be".

I already had a very educated idea of what we were in for. This was most likely extensive small cell carcinoma that has metastasized to other parts of the body. It is a very aggressive cancer and, if it is seen in both lungs, has usually spread to other organs as well such as the liver, brain, adrenals, and bones. Survival rate is slim at best, with combination chemotherapy and cranial irradiation as the treatment protocol. I kept this to myself for the time being, wanting him not to worry until the diagnosis was confirmed. I gave Bob the standard pep talk, telling him not to worry until we see the results of the scan and biopsy, and he chuckled to the punch line of some secret joke as he turned and left the office.

That night he trundled over to my house offering a casserole as payment for my consultation. I laughed and invited him in, made a salad as he set the table, and then we sat down to eat. After the usual pleasantries, he put his fork down and looked into my eyes, saying in a humorless tone, "Son, don't take up poker as a profession. The look on yer face when you popped the x-ray onto the viewer told me everything I need to know, so why don't you just level with me young man".

I was trapped by his frankness and compassionate for his right to know the truth, so after losing the staring contest, I took a breath and told him what I suspected: "Uncle, it looks like you *may* have cancer, and my guess is that it isn't going to be a simple fix. I could be wrong, and I hope I am, but after 12 years of training I think I have a pretty good idea of what is going on in your lungs. Once we determine the facts through these coming tests we can make a plan of attack and go from there". He tried to press me for more, but I was able to calm his anxiety by reminding him that not much could be determined by a single x-ray. The CT scan was scheduled for the next morning, I reminded him, and the results would be quickly forthcoming. We agreed to wait until there was more information before discussing it further.

The CT scan confirmed the need for a biopsy, which we had done right there that same day. More tests would come soon to determine the extent of the metastasis. Uncle Bob was stoic and contemplative, and had understandably lost his sense of humor. As the days passed and the various results came in, I took the time to level with my uncle about the difficulty of the battle he was in. He asked what would happen in the next few months, and I explained about the Chemo and, if the cancer had traveled to his brain, the radiation treatments that would hopefully prolong his life and reduce his pain. He asked about the chances of survival, and I told him the truth: ten percent chance of surviving past one year. I added that an oncologist would certainly paint the picture brighter, but I owed it to him to be realistic.

Uncle Bob looked at me and told me he wasn't ready to give up and die yet, that he planned to beat the disease. He said he was willing to do whatever was needed to stay around and beat the odds. I told him I was too.

Treatment was started and over the next few months his condition progressively worsened. It soon became obvious to him that this was a battle he was not going to win. In our many daily chats, he confessed to me repeatedly that he was afraid of dying and of not knowing what lied beyond this life. As the days passed he upgraded this fear to being terrified, and with each new discussion he became more afraid and anxious. My uncle was suffering a torture that infected his mind as much as his body. I worked with him as compassionately and lovingly as I could to ease his fears and give him hope, but hope for what? My own near death experience, combined with the Ayhuasca journey in the jungle had altered my view of what happens after we die, but how could I convey that beautiful place to a man so terrified of death without seeming patronizing?

Then, as if it was answering my plea, the idea that changed everything came to me in a dream that night as I slept. Most dreams fade from our minds within an hour of our waking, but for me, this dream had jolted me with an energy and a clear memory of it for long after, and I was able to transcribe all the details. I started putting together a plan of what I would need in order to make the dream a reality, and asked Sally to place an ad in the newspaper for another research assistant with a very specific set of abilities. I also needed to find a chemist, so I made inquiries at the University of Oregon and Oregon State University to see if there was someone the faculty could recommend.

Within a few hours I got a reply from Dr. Sandra Logan from the chemistry department at Oregon State, who said she had a couple of very talented graduate students who might fit with my requirements. I emailed both of them asking for their interest in my project and if they could interview right away.

That evening I brought my idea to my uncle. As we sat on his front porch I hesitated for a moment as I gathered my thoughts and courage, then told him my idea.

“Uncle Bob, I’ve been thinking a lot lately about your fear of what lies beyond after we leave this earth, and I wanted to share something with you. When I was traveling before college I had an accident, as you know, but what I never shared with anyone is that while I was unconscious I had experienced what’s known as a near death experience. I believe I was clinically deceased and then returned to the realm of the living a few moments later. In that time away from my body I strongly believe

I had a taste of what it's like after we pass on. What I experienced was truly wonderful. I was filled with a sense of love and of being back in my original home, in some comforting yet indefinable way.

"After Olivia died, I spent time in the Amazon basin, and while there I had partaken in a shamanistic ceremony, which through their sacred medicines, gave me another, much clearer vision of where our spirit lives when it is not occupying our body. These two experiences together removed my fear of death and replaced that feeling with one of contentment, almost like a sort of anticipation, I guess.

"I was thinking of giving people like you a glimpse into what could be waiting for them as a way to relieve their anxiety and fears, and maybe even get them to possibly look forward to the transition. That said, I was hoping to work with you on this experiment and see if we could ease your fear of what lies beyond."

"Just how would you do that? I mean, what do you mean by 'working with me'? You know I'm up for anything, and I love and trust you, but what do you have in mind"?

I told him that I planned on putting him under anesthesia and slow his heart and respiration to a close to death-like state, then administer a medication that would open his mind to what lay beyond. The same medicine I took in the Amazon. After a few minutes I would bring him back to full consciousness. If everything went as I anticipated he should be able to recall his experiences to us, which we would record. He said: “It sounds like you’re going to give me a Psychedelic trip, not an insight into what lies beyond. I don’t see a connection”.

I explained that DMT is a product of nature. It’s in many plants, as well as our body. We actually Produce it in our Pineal gland.

“Pineal gland”?

“The pineal gland is a weird organ. During the daytime it produces serotonin, which we know keeps us happy and buoyant, and at night the serotonin gets converted into melatonin, which helps us to relax and sleep. The gland also produces DMT, which is converted from serotonin because they’re similar; there is an enzyme that converts it. The gland is part of the brain’s structure, situated just outside but still within the spinal fluid, in a cavity. Why would we have an extremely strong psychedelic substance being produced in our brains? What is its ordinary function in humans? It’s not very well understood, but it is my belief that it is the bridge within us which brings us to the other side. Those who have taken DMT describe their experiences as the preview to what lies beyond.

"Dr. Rick Strassman, who did extensive clinical research on the substance suggested that what happens in a near death experience is that the pineal gland releases its store of DMT into the brain and it is this chemical release that gives the experience. My theory is that the purpose of DMT is to help people transition from a living state to a post-death state, whatever that might be. That's a quite ambitious speculation from a scientific perspective, but maybe with your help we can find the truth. People report having experiences of conciseness and conscious awareness even when there's no apparent brain activity. When the heart stops beating for about 20-30 seconds, all brain activity that we can detect stops because the brain is starved of oxygen.

"In theatre operations when people have had their heart stopped and the blood drained from their brain, many have reported conscious experience throughout the whole procedure, sometimes lasting an hour or more, during which there is no brain activity that we can detect. This challenges the notion that you can have no consciousness experience without brain activity. Now we can't be certain that they're actually having a consciousness experience, because we can't be sure it's taking place at the same time as their brain stops working. They could just conflate the experience when they recover brain activity. Or it could be that there are some parts of the brain that we can't detect which still remain active.

"We're in a position to experimentally put people into a death state and see what happens", I said.

Bob tentatively replied that he would like to see for himself what was waiting for us beyond the veil of our human existence, and if it could help me in some way as well then he was all in. By the end of our conversation he said he was indeed interested, and since he had nothing to lose, he said we should plan the experiment for as soon as possible. Now I just needed to get a hold of some DMT.

I interviewed the chemists in the following two days. After explaining what I needed from them, one backed out for moral reasons, saying she felt we were breaking the law of the bible. The other candidate, Maxwell Lazarus, seemed to be a perfect fit. Ambitious, bright, and enthusiastic, with heaps of confidence, he thought nothing of making compounds that might possibly be listed on the DEA's list of schedule 1 drugs. He was comfortable adding an extra atom somewhere in the molecular chain to alter the formula enough to classify it into what I wanted as a new substance without rendering unwanted effects. As I went further into the discussion of what I wanted to do, Maxwell became more and more enthusiastic to the point that when I explained the administration of the medicinal cocktail, he offered to be the first guinea-pig of our experiment.

Volunteers would be needed, and his age and apparent excellent health would be ideal for playing with a combination of potentially lethal drugs. He hired himself on the spot, asking when he could officially start.

He quickly replied to his own question: "I'm not doing anything this afternoon, so if you want, I guess I could get started right away. We do have to set up the lab and do a mess of shopping, you know. If you don't mind, Doctor, let me take charge of the chemistry lab. I'll pick up everything we'll need for the manufacture of the medicines. Believe me, I've got this down. I know all the supply houses around here and exactly what we need. The most difficult part will be getting the raw plant ingredients we'll need for the chemical extraction, and that part really isn't that hard at all. I've got sources".

I had a feeling he had dabbled in some extra-curricular manufacturing of psychedelics during his term in grad school. His long hair and casual attire (he had come to the interview in hiking boots, jeans, and a denim shirt) spoke of his empathy for the counter-culture. I was fine with that, as long as he performed his work here with precision and maturity.

Within three days he had transformed the garage into our laboratory. Max made five trips into Portland and Salem for supplies, and I could tell in the order of his purchases that he was a master of detail. That same afternoon a van pulled up to our clinic and two men got out asking Sally to lead them to the garage. Within three hours they had nailed into the perimeter of the room a four-inch baseboard and then covered the concrete floor with a new vinyl floor, which lapped over the baseboards, giving us a floor that could contain a large spill. The following morning, he came with his purchase of tables and shelving, as well as a day laborer to install the latter to his specifications. Off he went again to buy the larger items, then the more fragile ones, and then two trips for chemicals.

Around six-o-clock in the morning the following day I was awakened to the sound of a truck coming up the ranch's driveway. Maxwell was coming to start work. As he got out of his vehicle I shouted out of my window in the other house that he was early, we didn't usually start until 8:00, and for him to come over and get the keys to the clinic.

An hour later I brought him a cup of coffee, my excuse to peek in on the early riser. He was on his computer writing furiously as a large pot of something was simmering in the far corner of the lab. He thanked me and told me he had purchased a coffee maker as part of the lab inventory, so future trips wouldn't be necessary. He handed me an itemized invoice for the sum of his purchases, which came to just over $22,000.

"I put it all on my credit card. Wanted to expedite things so we could have some working compounds as soon as possible. I'd really appreciate it if you could get Sally to cut me a check this week. I don't want to have this go past due, OK"? I agreed. He continued: "You said you wanted to replicate the medicinal properties of the Ayhuasca medicine, which is mostly DMT. I can deliver a small batch of pure DMT within a week, and we can go from there. I want to suggest, since you've already been on an Ayhuasca trip, that we give you a dose of pure DMT so you can compare the experiences. If it is the same, we can add an atom of something inert and have a beta compound. If not, we can add whatever we need to get the same type of trip, sound good to you'? I told him it all sounded fine, except for the label of 'trip'. I suggested we call it a transition, as what we were trying to do was to replicate the transition from life to death.

A week later Max had his batch of DMT ready for our first volunteer, me. In my clinic's trauma room Jenny inserted an IV needle into my arm and hooked me up to a saline drip. She connected me to the vital signs monitor and we were ready to go. Eric turned on his video camera and audio recorder. Max observed from the far corner of the room. He told me to relax and close my eyes, and for the rest to be as quiet as possible. Then Jenny injected 6 mg of DMT into the IV. She counted as each mg left the syringe, and by the time I heard her say 'six', I was starting my journey into what lays beyond our world. This was officially our first clinical experiment. I was able to verbally describe what I was 'seeing' and experiencing as I flew through the heavens and made the acquaintance of indescribable beings. If anything, this was a clearer and sharper version of my experience in the jungle, and I commented aloud that if I wasn't being such a Doctor right now I would surely be able to go much farther out into the universe, that I was "holding myself back in the name of science".

After 45 minutes the medicine had run its course and I was back in the default world. This was a moment of catalytic progress. The all-important key ingredient had been made. It worked beautifully, bringing me to a place I longed to return to, a place of welcoming love that felt eternal, and leaving me afterwards in a state of enlightenment. I glowed for weeks with the knowledge I had gained on that voyage.

In a nutshell, this is what I got from my experience: We are all shards of the same whole that we can never really separate from. We are connected to everything and everything is us, and at the same time we are both ourselves and the other. We are always protected and loved, and there is nothing to fear.

A couple of days later we were ready to try the full 'near death' cocktail on Max. I asked him to try and describe his experience while he was under in the hope that I would be able to fine tune the anesthetic vapor to a point where we could get him to actually talk through his transition. Since there was no surgery or other reason to keep him deeply anesthetized I would vary the dose to see if there was a difference in his experience. We rigged him up to the drip and monitors and brought him into a state of unconsciousness. A few minutes later, when I was certain he was responding well to the anesthetic, Jenny administered the DMT. A wide, toothy smile came to his face as he started his journey. As he was under the influence of the medicines, I was monitoring his vital signs and brain activity closely. Heart rate slowed significantly and brain activity quadrupled, and then went off the graph. I decreased the flow of popofol and his lips started slightly moving. He was trying to talk. Another 5% less flow and now the words were coming out softly. Another fine adjustment of the anesthetic and his words were decipherable, yet a pin poked into his wrist yielded no response; the dose was perfect.

Eric moved in closer with his Zoom audio recorder

in order to catch the nuances of Max's narrative. We could make out about half of what he was saying, and at times he appeared to be asking questions. Then he would laugh, and once let out a joyful woo-hoo. The narrative continued until the medicine had worn off and his decreased brain activity announced the trip was over. I brought him out of his unconscious state. After awakening and regaining his senses he told us the story of his experience.

He too had felt the enlightenment and had been accompanied by the beings, which he described as angel-like, but without visible wings. "I knew they were there, though, 'cause they flew around with me". He continued: "I met God, at least it felt authoritative like God would. I asked her if I was dead and she laughed, saying that death is an invention of where you came from. Even though you feel death, you don't die, you just move elsewhere.

I asked Max why he referred to God as a she. He replied that the image he saw was a divine feminine, and she looked like the Good Witch from the wizard of Oz, but, he added, he could have brought that with him because he was watching that movie with his niece the night before. Either way, the presence was feminine for sure.

We both rated our experiences of transition and his

was much, much more profound than mine. There were a number of duplications, but he went way further than I and had made many more connections. We calculated that his experience was about eight times more revealing than mine. This we concluded must have been due to the presence of the near-death state induced by the propofol. Our hypothesis was that the less in this life we are, the more help we have from beyond.

I went to my uncle and told him we were ready to show him what was out there beyond this life. I explained briefly that Max and I had volunteered for the procedure and both came out of it with clear memories of our amazing experiences. He begged for details but I told him that I didn't want to influence his experience. It was important for him to give me his first-hand account and then we could compare notes. He agreed, although with trepidation. I assured him that there was nothing for him to be afraid of.

When we were ready for our first actual patient, uncle Bob, I was as nervous as he. This was a real cancer victim with perhaps a month or two to live, and he was someone I loved dearly. I didn't want to mess this up for either of us. Eric said he wanted to try a throat microphone, like what pilots of small planes wear, and we strapped one to Bob's neck. It worked well, but not perfectly, so Eric would use the Zoom recorder as well.

When Bob was comfortable, Jenny went to work poking him and wiring him up. He said he felt like an astronaut with all the contraptions hooked to him. I held his hand and told him we were all here for him and proud of him for his bravery, that we were about to start the procedure, and I would see him in an hour or so. Jenny started the drip and I attached the mask that delivered the anesthesia. He was brought to unconsciousness and left there for a few minutes. Everything was well. I ordered the DMT to be administered and then watched as his brain activity soared. We knew where he was and I hoped it was as good for him as it had been for Max and me. He tried to talk but was only mumbling, so I turned up the vapor to bring him deeper into his sleep. We waited with anxiety and great anticipation as his voyage ran its course. Finally, his brain activity slowed to normal and I started to bring him out of his slumber. All vital signs returned to normal, except for a more rapid heartbeat, which eventually subsided.

A few minutes later he regained consciousness and as soon as he was able to speak he excitedly told us the most amazing story:

"I fell into a deep and bright white, yet dark warmth, where I felt this overwhelming sense of love permeating me from outside. Then it felt as if I was being lifted up and started to fly, straight up into the night sky, where I was joined by angels, which seemed like miniature people, running dancing and flying next to me as I learned this new craft of flight. Stars were all around in the distance, and their movement told me of my velocity and course. I felt nothing of my body unless I wanted to reconnect with it, but there was no need. It was an anchor, so I left it way behind as I explored this strange new place.

"I saw what looked like fuzzy bubbles here and there along the way, and realized they were other travelers like me, exploring this new space. There were others as well, who were more solidly formed, who gave me the message of their having been out there longer than the fuzzy ones.

"The guides were still travelling with me, not as protectors though, as I now felt invincible, but more as playful friends escorting me on my journey. After a while I came to what at first looked like a cluster of stars, like a nebula, and as I approached it changed form into what felt like a teacher or an angel, so I asked it a question.

"What question, uncle?" I asked.

"I asked it what this place was, and why am I here. it answered me in the most comforting voice imaginable, and smiled brilliantly as it 'spoke', saying: ' this is everything, and you are here because you belong here. This is home.

"After that, I felt like I wanted to return to you and tell you about how wonderful this place is. Is this where we go when we die, Mike?"

"I really don't know, Uncle. Some say it is, I believe it is, but I think I'm going to have to do some research to see what I can find out. I understand that the great mystics of all the world's religions say that that's what it's like, that we are all interconnected; we are all one being or intelligence, that the 'I' is not our body or our mind, it's our consciousness. I don't know what they mean by that, but it seems to tie in with what you just experienced."

We all applauded each other. His tale was remarkably similar to Max's and he came through the journey and the medications without any ill effects. He was ecstatic and glowing, Happier than I had seen him in months, maybe years. Sally brought out a bottle of champagne and glasses, and we shared a toast to our discovery and the success of uncle Bob's transition.

Later I asked him about the voice, why he had said, "it". He said that he felt that was the natural sense he got. The voice was more than masculine or feminine, it was every voice.

In researching people's near death experiences, I noted that many of the descriptions were very similar to what my uncle had described, some describing a white light instead of stars, and some describing an escort of angels instead of primal animals, but generally the same overall feeling of seeing a light or other entity which drew the person into a universe filled with love, friends and family, welcoming beings, etc.

The mystics of all religions are the highest echelon of the learned contemplative thinkers who know the secrets of their ancient texts and have often experienced spiritual enlightenment. Their consensus is that we are all interconnected with everything, are all parts of the whole as well as the whole itself, that "we" are not our bodies or our minds, but the intangible spark that animates us.

I found a beautiful quote from an Indian Guru that captured the essence of that truth wonderfully: "What are our bodies, but a warehouse of eaten food? What are our minds but a warehouse of memories and imagination? After those are gone, what is left? That *other* is the spiritual truth of who we are".

A few weeks later, after sharing with my uncle all I had discovered, he told me he was ready to go out there for good. He was growing terribly weak and pale, passing blood and in great pain. He said he was dreaming of the place he visited earlier, was exhausted in his fight with death, and was fully ready to go. In my love and compassion for him I wanted to help, but was terrified of what might be if I made a mistake, if I was committing an illegal act, and if I were wrong and these reports were all just coincidental hallucinations of people wanting to believe in a life after death.

But my uncle was not some religious follower, nor did he previously believe in anything after death except a big sleep. Now he wanted to quit this life, to rid himself of the excess baggage of his body, now filled with the belief there was something else after he left it behind. He was in tears from the pain, and maybe also from the futility of existing in a life with no future. He reminded me that assisted death was legal in Oregon, and that we had filled out the proper forms two weeks before his first transition, so there was no need for *me* to worry. I thought of Olivia's last request too. I knew that I had failed her, and that she was better off after her passing. I couldn't make that same mistake with my dear uncle.

We decided to do the ultimate experiment. I was able to monitor everything he would feel as he went through the transition from living to not. We talked about what we both wanted to accomplish in this experiment, as now he wanted me to share vicariously in his adventure and document whatever I could of his transition and voyage. My trepidation was gone, turning into a feeling of anxious excitement, like that of an engineer at mission control as an astronaut sits atop the spaceship he helped design.

On the chosen day I called in my research assistants and we started the process, connecting IVs, monitors, and recording devices, both video and audio. Uncle asked to be off his pain meds until he was ready to go, so he could be reminded of his situation and be clear headed while setting his mental house in order.

We were ready, and so was he, so the process was put in motion. The law in Oregon states that the patient must self-administer any lethal medication. He reached out to the dose metering machine and pushed the button which would start the flow of the medicines to the IV line. I asked him to communicate aloud all he could describe once he started his final transition to the other side so we could record it all. We set up baselines of his vital signs, then started recording. On his request, I administered the anesthetic to sooth him, keeping him teetering lightly above unconsciousness, then a moment later the machine started the dose of the brain stimulator. I asked him to describe his experience as he could, and soon he was chatting away about the space he was in that exuded love, the guides, and the astral voyage. As he was in that place, he was still aware of the anchor of his living holding him back from his full potential.

“Mikey, I'm ready. Cut me loose now. And in case I won't be able to say it later, thank you.” The voice was sleepy and faint, yet perfectly clear and sharp.

I slowly administered the lethal injection, 300 mg of Nembutal, into the IV and watched the monitor while letting him know I had done it for him.

“You’re flying solo now, Mr. Brown, Godspeed”, said Max.

Bob kept talking as his heart rate slowed to zero. Brain activity was at full speed, twenty times higher than a normal conscious mind. He seemed to sigh an excited 'Oohh', and smiled. His body was now dead except for the remaining brain activity. He was whispering, emotionally charged at what he was experiencing.

He described the feeling of his anchor chain being cut, and returning home to a place where he was connected to everything and being filled with a warmth like he had never known, with love everywhere and nothing to hold him back. He saw that his body was just a temporary home where he went to learn the lessons of imperfection, and now he was really home, wiser for that time. He started to exclaim something as his brain activity slowed quickly and faded away into total inactivity, with little flickers several times over the following few seconds, then nothing.

Those of us who were observing now looked at each other with a menagerie of emotions, which spanned grief, excitement, the thrill of discovery, and amazement. Jenny and Eric reverently disconnected my uncle and wrapped him in his favorite blanket. I sensed that he was not here, not in this body at all. What remained was his past home, which, from what he said at the end, made it sound more like a vacation rental than anything.

6.

I decided to gather more evidence and information on these experiences, and to see if there might be others willing to participate in this new clinical study I was about to undertake. There must be many people in the same state of fear and failing health as my uncle, and I sought to find them and ask them to let me work with them in their transition into death.

We placed ads on the Internet and contacted sympathetic doctors for likely candidates. A couple of weeks later we received a call from Martha, a 36-year-old woman with aggressive ovarian cancer. Her parents, who admonished her by saying the disease was God's way of punishing her deviant homosexual behavior, had compounded her trauma. I told her that there were no judgments here, just compassion and care. She got on a plane the next day with her partner, Alicia.

Martha was pale, with bruises over much of her body. The medical records indicated that the cancer had spread throughout her body and into the lungs, giving her the added burden of toting around an oxygen cylinder wherever she went. We got the pair settled in the guesthouse, and when they were ready, I shared with them the tapes of my uncle's voyage into the great beyond.

She listened with hunger, nodding as the descriptions touched her soul. She grasped Alicia's hands and at times seemed to be praying to herself. Her pain wasn't as physical as much as the emotional decimation wrought by her parents. Their words laid her heart to waste, she explained. To have that wall of absolute rejection and hate from the folks who she had loved and trusted more than anyone had been too much for her to bear. She said that there was no place for her in this world if her parents could be like that. She told me she was given two more months to live by her doctors, but she wanted to end the pain now, before it got unbearable.

We ran tests on her, giving that she was young and still not in her final days. She agreed to take the medicine in a gradually increasing dose, to see how it would affect her. A moderate dose produced visions of symbolic profundity, but did not bring her to the launching area to the other world. She said it reminded her of a psychedelic experience she had had before, but this time he imagery was intimately powerful to her. It was as if she were in the presence of God, she reported. Her audio recording was difficult to distinguish, so we worked with her on communication while “under the influence”. It seemed that she needed to feel total trust in order to speak while medicated. Here her partner came into the picture. Instead of me asking her what she was going through, Alicia held her hand and caressed her face and body while soothingly asking her love all the questions we had prepared for the test.

When she had recovered from the experiment, we watched the video together and she was able to elucidate on what she had been trying to say while under the influence.

She said that she was ready for the final experiment, the combination with the anesthetic and the gradual exit from this world. Martha mat have been ready for it, but Alicia begged her to wait, and we wanted to let her body recover from the effects of the administered drugs. The couple spent the next three days as our guests making preparations and saying their farewells.

When the time came, the ladies were valiant and as ready as they could be for something so final. We had Martha find a comfortable space on her bed as we started to connect the monitoring wires and IV needles. She was squirmy and restless, and called to me for a change of her situation. She asked for another bed, one big enough that she and Alicia could hold each other through the transition. We accommodated them, of course, and her partner lay behind her as they spooned, whispering comforting things to Martha as we started the procedure.

Alicia got her talking right away, and she whispered with amazing clarity of each detail of her voyage, starting with the sensation of warmth from the medicines entering her blood, to her awakening, and then to the start of her crossing into the other realm of pure consciousness.

She described a white light that came for her, and then encompassed her as she felt a powerful uptake of love and acceptance. Someone was there to welcome her, not someone in a body, but someone who was of pure light. As she went with the being, she saw on the periphery many beings she felt she had known before. It seemed to her as if they were welcoming her back to their family.

She had no fear, and felt that this was home to her soul. The pain of her family's emotional brutality was gone, and she felt nothing but love from and for everything. Everything is here, she said. I am a part of everything and it is a part of me. I feel like a perfect being, like pure energy floating through the cosmos.

We were stretching the limits of her life now,

bringing her heart rate to less than ten beats per minute while bumping the dose of brain stimulant a little more. Alicia continued to caress her as we watched and recorded everything. I quietly signaled one of my assistants to give her a little bump more of the stimulant, and before she could move, Martha said: “that won't be necessary, doc. I'm already where I need to be. I am fully connected now. There are a billion beings here with me, and they all love me as I do them. The place is an infinite white dome with beings of pure energy everywhere, like stars in reverse. They are waiting for me to let go of there. It's all so beautiful and perfect. Thank you, doctor. Tell my love not to be sad, don't fear for me Alli, I'm here with you. I'm right here with you. I can't go on until you release me, doctor. I'm ready to go on. Goodbye”.

She said no more, thought she was smiling and seemed to be humming. Her brain activity was starting to come down, as if she were trying to turn it off. I administered the Nembutal and in a minute she had ceased to be what we define as alive. Alicia looked at me for confirmation of what had just passed, and I gave her a somber nod. She whispered to me and asked if we could leave them alone for a while. I acceded and covered them with a blanket before leaving the lab with my assistants.

A while later Alicia came out and thanked me for

my strength and compassion. She was sniffling and her eyes were red from her tears, but she told me she was glad it was done. "The tears are only my ego's way of coping with the loss. How can anyone begrudge her the chance to be in the beautiful place she's in now? It sounded so...perfect". I could not disagree whatsoever.

I invited Alicia to review the data with us, but she declined, saying that she already got the message. Before she left the ranch, I asked her to keep her experiences here in the strictest of confidence, as there were many issues that we weren't ready to broadcast yet, and, of course, there may well be legal issues here that Martha's waivers and releases couldn't possibly protect us from. She said she would never tell a soul what had transpired here, and that what we were doing was going to change the way the world thought about death.

Because of her physical strength, we were able to keep Martha on the brink of death for fourteen minutes. Her account was an amazing tale, which confirmed and greatly enhanced the data my uncle had left us with. The biggest difference was the 'white light' versus the 'comforting darkness'. I took it to be a perceptive difference between the subjects. Aside from the illumination issue, the rest of the stories had many similarities, including guides, feelings of love, perfection, contentment, beings all around them, and the sense of being 'home'.

Over the next year, we assisted 86 other subjects who were all, to some degree, at death's door. They came once we had received and reviewed letters from their doctors, their medical records, and all of their releases, waivers, and consent forms. We helped them to see the experience as something to be welcomed, that death was an important part of our earthly journey, and that what they were doing was giving themselves an option of quality over quantity should they decide to opt out early.

We gained volumes of information on the transition from the recordings and scientific data we had amassed. We found that the background and upbringing of the subject led them to translate what they experienced to fit with their taught beliefs, such as the Evangelical Christian who saw Jesus as the guide, and Saint Peter at “the gates”, followed by an impending interview with God himself, which he described as he was taught as a child; A white haired and bearded white man of advanced years, sitting on a throne.

Many patients saw friends and family members waiting for them and there to help them with their transition. The friends and families of these patients appeared healthy and filled with happiness.

One patient, DJ, who had confessed in our pre-procedure interview prior to that he was quite the larcenist and swindler who had made his way through life at the expense of the trust of others, spoke during his transition of a ride that was filled with demons and little gnome-like beings that were taunting him during the primary phase of his journey, saying that they were yelling incoherently with ire at him, but they were then left behind (actually about two minutes), and the feeling of love pervaded his "body" (he was the only person to describe having a body on the other side).

Elizabeth, a 58-year-old woman who had been in a powerful automobile accident, was disfigured and in a great deal of pain. Her back was broken, along with both arms and one leg, and she had lost the use of her kidneys due to the trauma of the accident. She had been through several hospitals in the fifteen months since the accident, and was now losing her fight to stay alive because of her internal organ damage. She was, if one were to imagine her before the accident, a very beautiful and well-tended woman who took great pride in her looks. The car accident had shattered her self-confidence and self-image along with her body. She confessed repeatedly to being worried more than anything about how she would look on the other side.

I explained to her that there were no bodies after death, as far as we had observed (this was prior to DJ, chronologically), and that no one had mentioned anything of physical beauty. She couldn't grasp the concept, no matter how much I tried to explain, so I ended up telling her that I believed that she would not carry her scars with her to the other side, because those who have been through our transition described the people they had seen on the other side as being beautiful and healthy. I read to her a condensation of the religious beliefs, which said we would all be returned to our perfect state, which she was able to hold onto as an assurance that she would be beautiful once again.

During her transition procedure, she forgot to narrate her moment by moment experiences, as we had asked her. Instead, we were witness to her conversation with her guides. When she started the transition, which I labeled the primary phase, she began her journey by crying several times over that she didn't want to die (though she had already pressed the 'start' button ten minutes earlier). Then she addressed another being and had a conversation, of which we were able to record her side. Here is a portion of the transcript:

"I don't want to die! I don't want to die..." (twitching, with fear on her face, tears from the left eye).

"What do you mean?"

"I do feel it... It's not so scary" (muscles of the neck

trying to move her head, as if trying to look around)

"Oh my! Yes, it is. I do feel it" (smiling)

"Let me try... Yes, it is perfect, I am perfect...."

"This is all so wonderful... Thank you for helping me..."

"I am home again... (here she faded into humming and speaking too incoherently for us to decipher, but her fear had been replaced with a serene sense of joy which flooded over her face).

In trying to make the transition as comfortable for our patients and their families, we shared as much of our information as was needed to put all involved at ease. Eric produced a video which gave them a good idea of what to expect during the time of transition. In it were the words of those who had gone before, complete with the visuals of what we were doing with the patient as they were giving us their descriptions. Of course, there were always a hundred questions to be answered, but by the end of our interview, patients were usually clear in what to expect.

The procedures all went well, meaning that all the patients died giving some degree of description of their experiences before passing into death and before passing they all appeared relaxed and content. Often, however, the family members they left behind were grieving greatly, and to accommodate them I hired Diane, a counselor to aid with the patients' family in coping and dealing with the voluntary passing of their loved one.

7.

Despite our best efforts to make everything clear, comfort all parties involved in every way, and have the families sign a confidentiality agreement, the inevitable was bound to happen sooner or later. About two years into our program came a phone call from a reporter at the Portland Mercury newspaper wanting to talk to me about what we were doing up here. Naively believing in journalistic integrity and wanting to work in a transparent and open light, I took a proactive approach and quickly called her back, inviting her to come up and see for herself.

We prepared for the interview with a crash refresher in all we had accomplished at the clinic, lest she would want to interview anyone other than me.

When she arrived, I gave her the tour and answered her questions with candor. She heard the tapes and saw the video, and was then, as with everyone else, filled with even more questions about the process. She wanted to focus on the drugs we used, and I told her that they were proprietary (Max had by now synthesized the new molecule) and only used for our trial to get the patients to describe their experiences, and that there were volumes of evidence illustrating similar descriptions from many people who had near death experiences prior to our little undertaking (no pun intended).

I tried to explain that we really didn't want any publicity for what we were doing, not that we were trying to hide our work or anything like that, but to keep the investigation of death from turning into a circus. She said she understood, with a smirk for emphasis. I just hoped the smirk was not directed at me for suggesting the media might get out of control.

At the end of the interview I once again emphasized that our work here was to glimpse into the future, as it were, and our aim is to help people to understand what lies ahead once they will meet their earthly end, and from what our research has shown, the transition is nothing to be feared. Watching her drive away, my intuition let me know that something with this reporter was not right, but by then there was no way to retract what we had shown her. We knew the truth and our scientific methods were sound, so there was nothing to fear.

When the story came out, it had placed the emphasis on all the wrong things, describing our work as giving psychedelic trips to the terminally ill so they could feel joyful about leaving their bodies behind. That we had invented a new pharma-cocktail which got the dying happy, and that we were operating an assisted suicide farm here on the ranch. Her story was titillating, but far from accurate. So much for journalistic integrity.

The story was picked up by the wire services and went viral world-wide. We were immediately inundated with requests for interviews, uninvited people were knocking at our doors, and a few picketers started marching around at the entrance to the ranch. We were compelled to hire security guards and a public relations firm to deal with the fallout. Their task was to keep the press away from us and make up quick "press kits" to satisfy all the requests from around the world giving our perspective of the work we were doing. In retrospect, it may have been better to have changed our phone number and let it die naturally (pun intended).

The circus had begun. First came the media, wanting to find out more than there was, so they could 'scoop' the competition, then came the madmen and fools, either wanting to go on the ultimate ride, as if this were an amusement park or hippie commune, or the religious fundamentalists with their interpretations of how we were blaspheming and satanic and all the rest. We were condemned as charlatans and hucksters and fakers, but I guess that's only to be expected when we were removing the veil of this very taboo subject. I expected that organized religion would chime in soon as well, given that our research had so far dissolved their theories of a heaven, hell, and purgatory, with which they had manipulated folks for thousands of years.

Basically, ignoring the extremists, we were still confronted by many sincere people who just wanted to check out of this world for non-medical reasons. I told these folks in a press release that we were doing medical research, and only helping the terminally ill through a thorough vetting process beginning with their own Doctor, so then we were then attacked for being discriminatory and infirmist.

8.

Max and Eric came to me with a proposal: why not open our research to everyone and let them decide whether they want to check out or not. Eric said that if we kept this from the people then in effect we were playing God, deciding who lives and who dies. I retorted with the fact that the law in Oregon is extremely specific about treating only terminally ill patients, and even if we were to put that aside, there was no way I would be offering this discovery to more than those suffering terminal illnesses.

Max replied that I didn't have to, that we could easily manufacture a device that the user could have implanted subcutaneously and triggered with a remote device only he or she could access. He pulled from his pocket a cylinder the size of my small finger and said that something this size could hold everything needed for the device. He felt he could get it even smaller if we could concentrate the Nembutal or change it for something else, like cyanide.

Max went on to exclaim that people should be able to choose when they want to check out, especially since the post-death experience is so fantastic. A person who is ill and suffering can choose when they have had enough and simply enter a code on their smartphone and go to sleep. What's wrong with that? I replied that if anybody could get one of these devices, what would stop depressed teen-agers from opting out before they had matured through those crazy years? "And what exactly *is* that?", I asked, referring to the object he had in his hand.

"Oh, this is just a dummy. I've been looking into what it would take to produce a portable device small enough for self-administration. We could put the medicine in here and the micro pump would administer it through this aperture…"

"Let me see that, if you don't mind". He handed the cylinder to me. It was about a half inch in diameter, three inches long, with rounded ends. It was made of medical grade stainless steel and had been superbly machined. This was more than a dummy slug used to illustrate a hypothetical future situation. My intuition put me on alert and I felt the hairs on the back of my neck stand up. I saw a faint line about halfway down the cylinder and instinctively tried to unscrew it at the seam. It *was* threaded and came apart at my bidding. The shorter end was filled with something solid, a machine.

"What *is this*?", I asked again, this time holding up the solid half of the cylinder.

"Well, I guess it *is* a *little bit* more than a dummy. I've been working on the mechanical engineering of the delivery system, trying to see how manageable something like this could be. This part contains the radio receiver, battery, and delivery pump. Look, we've been doing two or three of these procedures per week, I assume mostly because of our limited space in trying to make the patients feel comfortable in the guest house. And like it or not, we've become a death clinic. That's why people come here now. I think most of them would prefer to be at home with their friends and family in the comfort of an environment they are completely familiar with, instead of in some clinical environment, no matter how homey we try to make the suicide room."

He argued that we needed to consider the idea of total access much more deeply, but at least the technology for delivery to whoever wanted an implant already exists. We could develop the device and continue the debate at the same time, but the important point was to at least give the elderly and terminally ill the freedom to check out when and where they wanted; that a previously made appointment that gives the patient a specific time in a hospital bed was one thing, but the freedom to wait until all was in order in their life before their final moments could be achieved with a portable device.

"How long have you been working on this?", I asked, handing the device back to him.

"Long enough to see its potential and get this far with a prototype. There are countries where suicide is legal, and in places like Japan, sometimes considered honorable. With your discovery, Doctor, the paradigm will surely shift in the very near future. We are all here to seek an end to the unnecessary suffering and pain in the terminally ill. Well, there are around 150,000 people dying every day, and many of them are dying after protracted and painful infirmities. We can't possibly hope to help a fraction of one percent of them in this clinic. A device like this is the practical answer. And of course, there are all kinds of technical, legal, and political bugs that will need to be worked out, but if we have a device ready, the issues will resolve themselves concretely, and not put off because the answers are an abstraction. And besides, if we don't bring this thing out into the marketplace, I guarantee you somebody else will."

I suddenly felt that my research and practice was starting to spin out of control. Between the publicity, the kooks, and now Max looking to capitalize on the procedure, I saw it all slipping out of my hands. The feeling of fear of being blamed for future suicides crept into my head. I left the office and went for a walk through the peach orchids.

And what about suicide? Why is it considered a taboo? When my wife died, where did the guilt and fear I manifested about her having a premature passing come from? That night I opened my laptop and discovered volumes on the history of taking one's life. I found out about the origin of the precedence for keeping access to suicide away from whomever wanted it. Apparently with the spread of Christianity, the promise of an afterlife filled with comfort and riches was enough to motivate the poor, sick, crippled, and especially the slaves of the wealthy and powerful to take their own lives. Wealthy slave owners petitioned Constantine and the early Christian church to penalize suicide by their forfeiture of their heavenly rewards if a slave were to take his own life. Religious dogma began to shift toward greater restriction of one's right to suicide over several hundreds of years, tightening in the middle ages to the point when, in 1670, Louis XIV of France, in collaboration with the church, made suicide punishable by public humiliation of the corpse and confiscation of the person's property.

In Roman times, if one wished to end his life, he needed only to apply to the Senate, and if his reasons were judged sound, he was given hemlock free of charge. During the Renaissance, the secular humanist perspective contrasted the church view. In 1516 Thomas More wrote in his essay *Utopia*, that a person afflicted with a disease can "free himself from this bitter life…since by death he will put an end not to enjoyment but to torture…it will be a pious and holy action".

The rift between the right to die and those who wish to restrict it has become secretly political, masked in the guise of religion. Although the consensus of most civil liberty associations is that "the idea of government making determinations about how one ends their life could be in certain circumstances considered cruel and unusual punishment, the state bows to the will of insurance and health care corporations. Insurance companies are not obligated to pay death benefits in suicide cases, and the health care industry enjoys greater profits with every day they prolong a terminally ill patients' life.

Morally I knew what we were doing was honorable, ethically it was correct, yet legally we were on dangerous ground. Until now we had been left alone on our farm in rural Oregon, but the potential for blame and punishment grew with every story published in the media. And if we were to make a device that offered the world's inhabitants their choice in when they ended their lives, we would certainly become a target.

9.

Eric suggested that it was time to make all of our results public, to publish the body of our work, and let people decide for themselves about what we had discovered. Publishing would add credibility instead of letting the press and the fanatics distort what we had worked so hard to achieve. He suggested I accept the invitations of the news networks and make a couple of appearances on the biggest shows to state the facts.

He was working on a detailed video that he said could easily grow into a full-length documentary, suggesting we post something on the internet that the general public could relate to. People were starving for information about this and we had an obligation to share our findings with the world. To that I agreed, suggesting we start with something short, and then continue working toward a more in-depth film. It was indeed time to set the record straight. He handed me a script and said he was prepared to film me as soon as I was ready.

I read the script right then and said I wanted to do some editing, but would be ready to shoot the next day. I was to do three on camera segments and a voice-over narrative for the rest. It was a well-crafted and textured illustration of what we had accomplished and discovered. Eric said he had done a video interview of my uncle Bob after he had taken his test transition, and with my permission would place parts of it in the video. I told him I'd look at his rough cut and give him my input then.

The meeting ended and I went home to contemplate the ideas of a portable device, publishing our results, and a documentary film. My team was filled with passion for our project and working tirelessly to share our discoveries. Ever since the first newspaper report appeared, we had been flooded with genuine requests for assistance in the deaths of terminally ill people. We were operating at full capacity, doing our pre-procedure interview with a patient in the morning, sending him home for the two-week waiting period, then performing one transition procedure in the afternoon. That was our limit; between the review of the patient's medical records and the extensive interview, inevitably followed by at least an hour of Q and A, the morning was shot.

Then, the actual transition procedure could be theoretically completed in about two hours, but was emotionally exhausting for my team. We always took as much time with the patient as he needed, and by the time he had expired we were occupied comforting his family (and ourselves).

The idea of being able to assist all who applied without actually having to perform the emotionally charged transition procedure sounded to me at the time like somewhat of a blessing. It grew difficult for me to see so many people die, despite their uniform bliss at what they were experiencing as they left their bodies. For an outside observer, death is still a dark thing that takes its toll every time we bear witness to it.

Within a week the video was shot and edited. We worked to avoid the sense of promoting our clinic, the practice of assisted suicide, or any moral, religious, or legal issues. Eric inserted a legal disclaimer in the clip's preface as well. We reviewed it over and over with my attorney and the PR firm we had hired. We uploaded it and sent out a press release, and within a week it had been viewed more than two million times. A month later it had 65,000,000 views.

It was time for me to go big with our discoveries. And the media loved it. Apparently, my personality and message were a great fit for the mainstream media.

People loved the message of looking forward to something better than living once we pass from this life. The well-worn religious dogma of eternal damnation and a judgmental God was being cast aside for the new doctrine of love and joy regardless of our past errors. Of course, the television based media is constrained by two huge factors, cramming everything someone has to say into a short two-minute long piece, and the need to make the story interesting and light, no matter what.

In that vein, all the interviews were filled with the same questions: how did I discover this secret, what motivated me, and was there a hell? And my answers were expected to be the same. Once I said in reply that from all I've seen, Hell, or heaven, for that matter, is present in what you do with your life right here on Earth. The interviewer redirected the question to ask if there was any scientific evidence of Hell and my original statement was edited out of the segment.

But the truth was now out there. I was becoming a minor celebrity, loved by the people and despised by the fringe. I didn't want publicity, but our findings were too important to not share with the world. My 15 minutes of fame spanned three weeks on the networks, until the next big thing came along, and I was relieved that my time in the spotlight had ended.

Shortly before my last interview, while I was in

New York, Max didn't show up to work. He vanished without so much as a phone call. Jenny noticed that the photographs of his family and dog were missing from his desk and wall, so she searched a bit and found a bunch of stuff had disappeared, including CD copies of our files. I called him, but he never picked up. Eric went to his house, but it was closed up tight and vacant. My assistant Anne, who had joined us six months earlier, decided to call his girlfriend, and was able to find out what had transpired. It seems that Max had decided to cash in on our work and make the technology available for anyone who wanted to get into the suicide business. He had partnered with a medical device manufacturing company in Stockholm Sweden, Senzime, to mass produce and distribute his self-administration device. Not only that, but he had been contacted by a publishing house with a contract to write a book; not so much a technical work, mind you, but a piece of easily digestible pulp designed for mass consumption.

10.

Within a month the Book, titled: “Beyond Death”, was being sold everywhere. Max did an admirable job compressing our research and results into 250 pages. I felt mildly betrayed by him, not so much because he was running with our discovery and making money on it, but more that he didn’t discuss it with the rest of the team. We had all worked together for two and a half years and were bound into a tight-knit team by our mutual experiences in the development of our discovery. Max’s sudden departure and decision to sell our work out seemed so unlike him, so out of place. But it was done so we let go of it and focused on the present.

Making the device available to physicians would take at least two years, probably more like three, given the FDA’s protocol for testing, but once approved, it might be accepted as an alternative to offer moribund patients for the few doctors who placed compassion over profit.

The following New Year's Day I got a call from a Dr. Lin, a coroner in Portland, asking if I could come to the city hospital and identify something for her. She avoided my inquiries and repeated that she would like me to come at my earliest convenience. I asked her if that meant *today,* and she replied that since she was obligated to be working on a national holiday, she would appreciate it if I could drop by if I wasn't too busy. Being a widower and deeply immersed in my career and discovery, I was indeed home alone that New Year's Day.

I arrived at the city morgue a couple of hours later. The medical examiner and her assistant took me to their autopsy room where a body was in the midst of his post-mortem examination. She nodded toward the body and picked up her notepad. She read: "Doctor Brown, this is the body of one Michael Schaffer, of Saskatchewan, Canada. 52 years old, white male. He was stopped this morning by the Oregon highway patrol for excessive speed on the interstate. Once he stopped his vehicle on the shoulder, he ran away from the car until the officer ordered him to stop running. According to the report, the subject turned back toward the officer and told him not to come near. He raised his arms as if to indicate he was unarmed and willing to cooperate. In his hand was a cell phone.

"As the officer tried to close the distance between the two, Mr. Schaffer stepped back and yelled at the highway patrolman to stay where he was. The subject then yelled that he didn't do anything wrong and just wanted to be left alone. The officer attempted to engage in a conversation as the subject appeared agitated and emotionally charged. Another OHP unit arrived on the scene, parking in front of the subject. As the officers tried to slowly close in on him, Mr. Schaffer yelled at them again to stay away, followed by the words: "don't make me do it".

"Then he apparently dialed his phone and within 30 seconds had feinted, falling to the ground. By the time the officers got to him, Mr. Schaffer was completely unconscious with a slow heartbeat. An ambulance was summoned. Paramedics registered his vital signs as weak yet stable. They transported him to this hospital and 12 minutes into the transport his heart and respiration stopped. Resuscitation was unsuccessful, obviously, and here he lays.

As she spoke my guts flipped and my body contracted an instant case of goosebumps.

"We've requested his health records from Canada. A search of his car produced two different prescription antidepressants and prescribed Hydrocodone, Zofran, Maxeran, dexamethasone, as well as THC concentrate.

Our autopsy revealed a few interesting findings, including one tiny, atrophic kidney and another large kidney smothered in cysts.

"Sounds like he may have been suffering from renal cell carcinoma", I numbly, automatically stated.

"So it seems", she replied. "We will have the blood and tissue tests back tomorrow. But the *most* interesting thing we found was this", and she handed me a baggie with Max's stainless steel cylinder in it.

I took the bag and held it up to the light, hoping to look as if I were examining its contents, trying to stall and collect my thoughts. How was this possible? Barely seven months ago Max had defected and supposedly sold his device to a Swedish company. They would need two years for approval and then another six months to start manufacturing, then there was distribution and all the rest, so this could not possibly have been from them. And in all that had transpired since our first conversation regarding the cylinder, I imagined that only a handful of people could have ever even seen this device…

"We found this object implanted between the adductor magnus and the adductor longus muscles of his right thigh. I believe it is a suicide device. It may have been triggered by whatever he sent from his phone. We can't do anything here today, but tomorrow when everyone is back at work we'll have this thing taken apart and analyzed.

"Doctor Brown, you have more knowledge in this area, that is, assisted death, than anyone else in the state, so I wanted to get your opinion of all this mess.

I handed her back the bag and looked at her, trying to come up with something singular to say. No words came, so after an awkward silent pause I said, "I need to sit down somewhere". She suggested her office, so I followed her there. As I walked I considered my role in this man's death. Was I responsible? He was likely suffering from depression and had been diagnosed with a devastating form of cancer. He sought, on his own, a way to end his life at a time he could choose. He must have gone to extraordinary lengths to find this device and get someone to implant it for him. My discovery of what lies beyond never condoned suicide, much less with the device Max had developed. Yet this reality fed my guilt for having opened the Pandora's box and revealing the secret of what lies beyond.

We sat across from each other at her desk. I started the conversation: "Nine months ago one of my assistants showed me a cylinder like the one you removed from the body in the other room. He said it was a prototype of an implant which the user would be able to operate remotely, via his cell phone or another device. He had designed a delivery system for the medicines we were using to assist our patients in exiting this life. He thought the masses should have the right to decide their end, no matter what. The story you told me a few minutes ago was exactly what my assistant saw in his vision for the future of this technology.

"A few weeks later he disappeared, taking his device and copies of all our research with him. We have not heard from him since. He wrote a popular book on our findings and from what I understand he contracted with a medical device company in Europe to start the process of testing and possibly marketing the device with the FDA and its counterparts in other countries across the globe. But that process would literally take years to complete, so I really have no idea where this device could possibly have come from. It was either supplied by my former employee directly, or he sold his technology to more than one company and the second one is not respecting the rules of the game. I imagine Max's prototypes would have cost him many thousands of dollars each, so he probably wouldn't have donated one to some Canadian man he didn't know. There must be an explanation.

Doctor Lin replied, "I'm sure the police will look for those answers, if they think it should be pursued. I'm still trying to get a handle on the potential problem we could have on our hands if this device becomes accessible to the general public.

"It appears that it already is", I injected. "This man, who lived 900 miles away from here in another country, was able to find access to it. He may have had it implanted anywhere between here and there, but he did have it done. I wouldn't be surprised if there is a network or online chat room that can give folks like him the answers he sought.

She and I discussed the issue and the man until we realized that we had exhausted the topic with the limited information we had at the moment. She agreed to let me know what the lab results and inquiries to Canada and with the police revealed. I offered to search the internet for clues as to where the device could have come from. I drove home wondering about Max and how radicalized he may have become.

I called Eric and told him about my New Year's visit to the morgue. He said he was shocked, but not surprised, as he and Max had had many discussions concerning our discovery. He said Max had indeed become zealous about sharing the technology and giving the freedom of choice to everyone. That's why he wrote the book, to get the word out to as many people as possible. He said he was going to churn the profits from the book into translations and distribution to people all over the world. Eric told me that Max had a vision of his implant being available and affordable to everyone. I replied that partnering with Senzime would preclude offering the device to everyone, as they were a company based on big profit, and they would surely concentrate on selling the devices for many thousands of dollars each to Doctors and institutions in America and Europe.

Then Eric said, "He was really passionate about free access. Maybe he used the front money from that company to set up his own factory elsewhere. You know, you can have almost anything manufactured discreetly in China, without a license or government oversight. There are brokers and facilitators all over the web willing to bridge the gap between design and manufacturing. I'll bet that's what he did, Doc".

Had I missed some key to Max's personality? It really didn't matter now, though the signs were present all along. He was a passionate individual whom, from the day I met him, showed his flexibility in regard to the law and societal norms. He had been instantly enthusiastic about manufacturing controlled substances without government permission, and although I was much less so, I was indeed complicent and quietly excited that he was of this type of free spirited entrepreneurial character. Without his willingness to bend the law we would not have been able to have made our discovery available to our patients.

When he vanished, it wasn't with malicious intent. He had only taken copies of what he needed, not originals. He could have caused harm to our clinic and the path we were on, but I believed that he truly wanted us to succeed. Upon reflection, he may have seen me as a father figure in our work, and his sudden departure may have been because he didn't want to be dissuaded from his passion, his plan to share his discovery with the world. We had butted heads on the subject enough to have established our positions, and he knew I wouldn't have agreed with his wanting to 'fast track' world-wide availability. And now, just a few months later, one of his devices was removed from the thigh of a cadaver in Portland. I wondered how many more of those things had been put into people, and who was deciding which patients were appropriate candidates. I imagine that with Max there was no vetting process other than the individual's desire to have control of when he or she wanted to end their time on Earth.

That evening I called a friend of mine who knew much more about the world of the web than most, certainly worlds more than me. His name was Raj Patel, and he worked for Oracle systems in San Francisco. He had been a hacker in his youth and because of his talent and success had been hired by many of the tech firms in the Silicon Valley, moving up the corporate ladder with each company. I couldn't pretend to know or understand his work, but I knew he would have answers for me.

I explained to Raj about the apparent suicide and the device Dr. Lin found, and about Max and his desire to make it available. In his reply, Raj told me that the internet was made up of many layers, like an onion. If one knew where to go, he could get to another, deeper layer. What I knew as the web was the top layer, called the clear web. This top layer consists of about 5% of the entire web structure. Then there are layers for corporations, for banking and governments and intelligence agencies, which is known as the deep web, and within that layer is located a sub web known as the dark web or dark net, where taboos are available and discussed, shared, and exchanged as if there were no boundaries nor legal constraints.

If I wanted to go deep, he suggested I create a VPN, a virtual private network, and a TOR account. TOR is an encrypted browser which can plunge into the depths of the deep web without risk. Despite the libertarian tenants of the deep web of freedom of expression, there are always those out there prowling for people to take advantage of. In the preservation of individual privacy there are sites down there offering everything from drugs to murder for hire.

He gave me a quick tutorial on how to set up a virtual private network account, which bounces my browsing activity to servers all over the world and makes my signature untraceable, and then connect it to a TOR account, which I found quite easily. He told me to make sure all other apps on my computer were closed, that I turned off Java Script when in the deep web, and to put tape on the camera and microphone of my computer so no one could capture my image or voice, and then I would be secure to browse the deep web at my leisure. I thanked him for his help and soon started my exploration of the underground internet.

After a day and a half of curiosity, shock, and amazement, I found what I was looking for. I could have gotten there much more quickly but was as distracted as a child with a new toy at what I found as I browsed. There was lots of news and images of war that would surely have been censored on regular channels of the media, marketplaces for anything one could ever hope to find, legal and otherwise. Most every book that had ever been banned had reappeared on the dark web, available instantly to whomever wanted them. It was an astonishing display of the ultimate freedom in a world without boundaries or rules.

There were forums on suicide, sites dedicated to assisting with suicide, and all manner of items for sale which would make taking one's own life easier. But then I came across the site called Better Than Living. It was a clone of all we had created in our research and what we had published both publicly and privately. There were videos of several of our patients passing from this life to the next, complete with subtitles for the hearing impaired. I was quoted throughout the site, as were passages from Max's book. The opening paragraph read as follows:

"Are you in pain, dying, or depressed? Is this life proving to be less than you had hoped for? If you are not filled with joy with who and where you are in this journey, you have the option of getting off this ride at the time of your choosing. Society has brainwashed us all for generations into thinking that there awaits a powerful judgement once we die that will either reward or punish us for what we did while on earth, but now Scientific discoveries have exposed that lie. We now know that what waits beyond is an existence much, much better than anything we could imagine. You deserve to be the master of your destiny and we are offering you the tools do take control of your life".

It was an impressive sales job, and after reading the rest of what the site had to offer, I could see how folks would find having the option of when to go irresistible. The site took the fear out of dying for whomever read the copy. The third paragraph introduced the device, now even smaller than the one that Dr. Lin had removed from the late Mr. Schaffer. This one was the diameter of a drinking straw and about five inches long, and was now made of plastic. Virtually undetectable in an airport TSA screening, and worked when the user entered a three-step code into their cell phone. It was guaranteed to work independently of network or cell phone service availability.

Most amazingly to me was the option for the buyer to purchase a self-installation kit, which included antiseptic scrub, a local anesthetic injection, and a sterilized, pre-loaded “syringe” which penetrated the skin and deposited the device in the sub cutaneous fat along the inner thigh, to bandages and adhesive for sealing the 4mm wound left by the syringe. The site claimed that almost anyone should be able to insert the device, and if they couldn’t reach their thigh, the abdomen would work as well. If they wanted immediate results, they could use the installation kit to get the device down the throat into the stomach, and had a 12-hour window to activate the “Gateway”, as it was called there. The quickest option was to use it as a suppository and activate it within 15 minutes.

It sold for less than $300.00, $400.00 with the self-installation kit. The price, as the site stated, was their cost of production, and if the purchaser could afford to do so, a sliding scale donation of 'profit', from 5% to 100% was requested. They also offered the option of donating more than the profit percentage as a way of giving the poorest of people access to the devices for less than cost. That was termed the 'scholarship' donation. The site claimed that in the hope of making the Gateway available to all, society would be able to break the chains of church and governmental control of our lives. With half the world earning a dollar a day, the price of the gateway is out of reach for so many, and a donation will give those less fortunate an opportunity to choose their destiny as well.

I felt Max's presence permeating the site. It was as if he and I were back in the clinic once again, debating the Impact of our discovery. I had little doubt that he was the owner of the site, so I sent him a message. I asked about his health and well-being, then went into my story of the device found on the dead Canadian, the report and curiosity of the coroner, and asked him how the business was doing. I wanted to reestablish a rapport with him and let him know I was not an adversary, and I wondered how many people were buying these devices. I submitted to the urge to buy a device, to see how the mechanism worked, and at checkout I made a generous contribution to the scholarship fund. The screen told me the shipment would be delivered in three weeks. No clue as to why the delay or from where it was being shipped.

I called Dr. Lin and told her briefly of what I found on line. Without mentioning the dark web, I told her of a site selling a device called a gateway that was much smaller than the one she had found, and it was sold with the kit to self-implant. When she wondered aloud about the quantity sold, I replied that I had written to the site administrator asking that precise question. She told me that the police had not been in touch, but she had gotten the medical file on the late Mr. S, and he was indeed suffering from cancer and depression, and from what she had read on his history, there was no clue as to why he was in Portland. From the implant scar, she deducted that the device had been implanted more than two months ago. And that was all she could put together on that case. It felt to me as if her curiosity was waning and she was ready to move on to more pressing matters, so I thanked her and we agreed to get in touch if any further news came to light, but I had the feeling that she was done with me and Mr. Schaffer.

11.

I checked my new anonymous email in the deep web every evening, hoping for a reply from Max, or whoever may have been running the Better than living site. On my fifth night, I opened my email account to find a reply:

"HELLO DRB, I AM SURPRIZED YOU FOUND ME HERE. I ALSO PROMOTE THE GATEWAY ON THE CLEAR WEB AND EVERYWHERE ELSE I CAN. THANK YOU FOR YOUR DONATION. IT WILL BE WELL USED.

"I AM WELL AND IN HIDING. THERE ARE MANY WHO WANT ME DISAPPEARED. THEY WILL BE COMING AFTER YOU SOON, OF THAT I HAVE NO DOUBT. WE HAVE REVEALED A VERY FRIGHTNING TRUTH THAT SOME BELIEVE SHOULD HAVE NEVER BEEN DISCLOSED, AND THEY ARE MOUNTING A CAMPAIGN TO DELETE US AND OUR FINDINGS IN THE HOPES OF GETTING THE GENIE BACK INTO THE BOTTLE. WE MUST PUSH BACK.

"WE HAVE SENT OUT MORE THAN 8 MILLION UNITS TO DATE AND AVERAGE 14,000 PER DAY DIRECT AND AS FAST AS THEY CAN BE PRODUCED. THIS IS MY GIFT TO SOCIETY. I AM SURPRIZED YOU HAVE HEARD ABOUT A USE, NOT SURPRIZED A GATEWAY WAS FOUND, AND NOT SURPRIZED THAT IT HAS NOT MADE THE PRESS. BUT THIS WILL REACH A TIPPING POINT AND THE DOGS WILL BE LET LOOSE BY THEIR ELITE MASTERS. OUR OFFERING WILL BE CONDEMNED AND WE WILL BE DEAMONIZED.

"PUT YOUR AFFAIRS IN ORDER AND PREPARE TO RUN. I WILL WARN YOU IF I CAN. GOOD LUCK, ML

I replied, trying to discover what he meant, but there was no response. I shared the note with my staff. We agreed that the threat to our peaceful clinic may indeed exist. On an instinctual impulse, I asked the crew to start separating the stuff we use often from the rest so we could pack and store the non-essential items for storage. We would streamline our operation in case we encountered problems.

I told Eric I wanted to make digital copies of all our files, with sensitive personal information redacted, so if the powers that be did try to confiscate our research, we could publish everything on the web via Wiki leaks, the deep web, and elsewhere. I never wanted to keep what we were doing proprietary, and if they were going to try and suppress the truth I would fight back. All of our formulas, experiments, records and the rest of what we had been compiling these past three years were copied onto flash drives and CDs.

Once we had gotten our operation streamlined and our data safe and secure there was nothing left to do but continue on with our work of assisting the terminally ill assuage their fears and make the transition into what lies beyond. I checked my anonymous email every night, kept sending (unanswered) messages to Max. I became worried about the authorities coming to shut us down a little more every day. Max's words of warning had ignited concerns within me, and the idea of going to prison for what we were doing grew more foreboding over time. I started searching for alternative sites outside of the United States.

12.

In my office were almost a thousand letters from prospective patients, the office email account had another two-thousand requests for our service. Almost all of these were from people genuinely in need of end-of-life palliative care. At six patients per week, we had been overwhelmed with requests since the original story was reported. Even with its skewed slant on the clinic, there were many people who saw what we were doing as important to their final time on Earth. our backlog of patients waiting for our service was now well beyond their average life expectancy.

How could we do more? How would we be able to assist these people with their transition without making them wait until it might be too late, or decide who we would treat and who didn't measure up? I had hired another worker, this time a psychologist, Richard, to deal exclusively with the constant, growing flow of requests from those asking for our help. At the end of his first week he met with me to discuss the emotional crisis the work was creating for him: first having to evaluate who was worthy (98% or so were truly terminal and afraid) and then how to prioritize them; what was the protocol: life expectancy, gravity of their fear of the beyond, or the amount of pain they were suffering? That was phase one, he said. Then came the truly soul-crushing task of calling those who were in the bottom 98% of the list and having to tell them that we were too *full*, too *booked up*, too *busy* to choose them.

He was in tears as he came to the realization that we had to find a way to expand and grow in some way that could not let these people down. They were at the end of their lives, without any hope except what we offered them. Life had kicked them down to the point of staring the grim reaper in the face as he sharpened his scythe, they were terrified of the future and we offered a beautiful, hopeful solution, and now it looked to them as if we were yanking that gift away in an act of absolute power they had no way of combatting. It was to many worse than their original terminal diagnosis-their last request for grace summarily denied by a voice on their phone.

I empathized instantly, my memory flashing back to the tremendous fear I saw in so many of my patients, and called a staff meeting for that evening. The crew were so tremendously valuable in insight that all the big questions were brought to the group for brainstorming. We all had a huge emotional stake in what we were doing. It had become a career of overwhelming ambivalence for us all. The polarity of what we did and felt was processed differently by each of us, so everyone's input was extremely valuable.

Richard started the meeting reiterating his dilemma. We all agreed that we were in an unsustainable and frankly unfair position-turning down patients in need was unconscionable. We were emotionally charged and determined to find resolution, but our options all seemed difficult at best: moving to a huge building and increasing staff, franchising to doctors throughout the state, doubling our current workload… there was no quick and uncomplicated fix that any of us could see.

And then Eric stood up, calm and confident, and posed his solution.

"You know Doc, I think we need to consider the past as a way to address the future. And it might make you uncomfortable to go there, but as I see it, what Max did could be our salvation, as it were. He is mass-producing the same thing we're offering here. You said he's making thousands of units per day, so he has the inventory to cover our needs. Imagine if we re-engineered our paradigm into a model of streamlining the intake process, inserting the devices, and having the patients return home to be with their loved ones when they are ready to make the transition, we could attend all those waiting people in short order".

He paused, having run his spontaneous idea to the end of its germination for the moment. The rest of us were silently processing it when Richard spoke up.

"You've all been on an emotional roller-coaster since you joined the clinic. This plan would remedy all of that quickly. Doctor Brown, you have more than two years of clinical trials with an impeccable record. Your work is being reviewed and proven safe to all involved. The Netherlands wants to adopt your method for their national assisted suicide program. You, as they say, have 'been there and done that'. You have succeeded at what you set out to do.

"For the entire staff, seeing so many people die after the long and intimate counseling process, and then the daily procedure of taking their lives, not to mention the disposal of the bodies and sharing the families' grief, the emotional toll to us is unsustainable. It's time to let go of the control and make the process less clinical and more in the hands of your patients.

I was at that moment hit with the most intimate image of my dear Olivia begging me to end her suffering. If she would have had an implanted device she could have been the one to decide when to say goodbye. She knew when she was ready to let go, and in my vision of her she looked at me and told me to do this. It was the right path. If she had been in control back then I wouldn't have been wracked with guilt for the next few years. A guilt which subsided when I assisted my uncle into death. What I felt now was the true liberation that so many would have. Even in the clinic, as much as we were trying to help them, the patients were on our schedule, not theirs.

We agreed that becoming a facilitator for home assisted transitions would be the best solution for all. It felt, as we left the meeting that night, that a pressing weight had been lifted from our collective souls. I asked the staff to work on new processing and counseling protocols while I would do my best to reach out to Max.

13.

That night I went to the dark web and wrote to Max describing our position and our intended path, hoping for an answer but in reality expecting none. I had no idea of our course if he didn't respond. I pressed SEND and stared at the monitor, entertaining the fantasy that he was out there somewhere glued to his inbox waiting for me to reach out to him. After a while I gave up and went to bed. I got up twice during the night to check for messages but saw none.

The next three days were the same, zero new messages. I told the staff we would meet that evening to discuss our situation. That afternoon, as I left the transition room after assisting another patient in passing from the Earthly realm, Eric was outside the door and grabbed my wrist, pulling me with him to the lab.

"Doc, you've got to see this"1 He whispered, obviously excited.

As we entered the lab he let go of me, closed the door, and said, "This afternoon a FedEx van pulled up to the clinic and unloaded 33 boxes marked Medical Supplies. Sally signed for them and I hauled them into the lab. The first three were regular inventory restock. The fourth box I opened was this one".

He pointed the box out, then as if I wasn't grasping what he was trying to anxiously convey (which I wasn't), He picked up the box and unfolded its top flaps to display its contents. I looked in and saw a bunch of clear plastic sealed bags with what looked like opaque plastic tubing within, as well as a folded paper which I guessed were instructions or a warranty for the plastic tubes. I took a pack from the box and looked blindly at it, still not focusing as I was emotionally still back in the transition room and dazed by Eric's excitement.

"Don't you see what these are?", he asked.

"I'm trying to fig…"

"They are the kits Doc! I haven't opened all the boxes, but there's a hundred per box and thirty boxes of them, so it looks like we are set to make history again."

I was speechless. Now the staff meeting would be a celebration. I needed to process the surprise and rethink the message for the meeting, so I went home to shower and decompress. As I opened the screen door at the front of my house, a large manila envelope fell onto my feet. Hand written, addressed to me, but lacking postage and a postmark. There was no return address either. My mailbox was at the entrance to the property, so someone had made a concerted effort to bring the package 200 yards up the driveway and place it behind my screen door.

The curiosity in me led me to the kitchen table, where I opened it to find a note from Max, though it was not signed.

IT'S A GREAT MOVE. SO GLAD YOU THOUGHT OF ME. I AM HERE TO HELP AS MUCH AS I CAN. ENCLOSED ARE THE CLINICAL SPECIFICATIONS ON THE MEDICINES THE CYLINDER WILL DISBURSE AND THE TIMING OF EACH. WHAT THE PATIENT CAN EXPECT IS ABOUT THE SAME AS YOUR PROCEEDURE, BUT THE 'JOURNEY' PHASE IS ABOUT 25% SHORTER.

YOU CAN PAY THE INVOICE TO THE ADDRESS ON THE SITE. MORE KITS WILL BE AVAILABLE AS YOU NEED THEM, BUT PLEASE BEWARE, THERE ARE POWERFUL FORCES WORKING RIGHT NOW TO SHUT YOU DOWN AND DISCREDIT YOUR WORK AS FRAUDULENT,

POSSIBLY WORSE.

PREPARE FOR THE WORST AND BE READY TO DISAPPEAR INSTANTLY. WE WILL DO OUR BEST TO GIVE YOU NOTICE.

GOOD LUCK.

I glanced over the paperwork, which was thorough and as detailed as I could have asked for. There were chemical compound breakdowns, dose metering notes, and detailed instructions designed for a medical professional to insert the device properly into the patient's thigh. It was technologically advanced yet elegant in its simplicity. An outpatient procedure which would take less than 15 minutes to implant and another 10 to program the patient's smartphone app with them.

At the meeting we went over the intake and counseling considerations and felt that a four-hour group session with private follow-up as needed could instruct the patients on what to expect and how to proceed when ready. We would work a morning and an afternoon shift. Those patients coming for counseling in the morning would be implanted that afternoon. Those counseled in the afternoon would return in the morning for their device.

I figured that between Jenny and I we could do about 50 procedures per day, and working six days a week we should be able to get caught up with demand in about 12 weeks. It seemed daunting but doable, and we wouldn't need to say no to any of the truly needy patients who were reaching out to us for help.

We tested the device on our next five clinic patients, explaining the updated procedure as if it was always done that way. It was a beautifully balanced delivery system and made our job less stressful. The anesthetic was based on an average weight, so the fine tuning of the patient to afford him the ability to describe his journey to his family was not available, though some were able to get to that place just as if I were dialing the flow myself. Others didn't speak, but the monitors showed the same results as our previous procedure. They all smiled as they went on to their end.

After that trial we started sending patients home to self-administer their activation code when they were ready. The counselors asked family members to follow up with us after their loved one passed, and as the calls came in over the next months the consensus was that the home experience, the when and where, had improved the patient's and family's experience.

14.

As with all things, this stage in our operation was but a brief lull before the next storm. We settled into a rhythm of processing our patients in an efficient and compassionate manner within six months. At first we worked long hours to make up for the backlog of qualified patients, but then after a few months we were able to downshift to one turn of patients a day.

The patients would arrive as a group at 8:00 in the morning for their introduction to the process and counseling. By noon we would start implanting devices. The 'guests' were able to enjoy a light lunch and stroll the grounds as they waited their turn with me or Jenny. Once we finished with them they had a refresher tutorial with Eric on how to activate their device once they were absolutely ready to make their transition. This meeting with Eric was private, without family present, which left the details of activation solely within the possession of the patient.

All paperwork was handled in advance via email. The patients were required to provide their medical records, a letter from their doctor, and letters from themselves and family members stating their situation and their reasons for wanting to terminate their life prematurely. This gave Diane and Richard an opportunity to vet the applicants and prepare for any patient who may have special needs.

We hired a caterer for the food service, and considering the state of the patients and their upcoming outpatient procedure, the food needed to be all organic, of the highest quality, and vegetarian. A landscape artist came in and converted the area surrounding the clinic into a spa-like setting with parking and paths and semi-private coves along the path for contemplation. And of course, there was a water feature at the center of it all, a beautiful Koi pond complete with falling waters and water lilies.

The flow of patients ended up averaging just over twenty per day, or a hundred every week. The collective stress of the staff was significantly diminished. We were no longer subjected to the daily testament of witnessing people die in our treatment room, the arrival of the undertaker to collect the body, and the inevitable grief and our consolation of the families.

I was finally able to schedule a vacation for everyone. I decided to close the place for three weeks and give the team an incredibly well-deserved break.

While visiting my parents in Los Angeles I got a call from a reporter form the New York Times, Nicholas Kristoff, asking for an interview. At first I declined, but then he told me that my research had taken many people into a clearer understanding of the process of dying and he wanted me to summarize what one could expect after death for a segment of society that got their information from outlets like his rather than YouTube and Facebook. He told me that statistically suicides were now more than seven times their rate of three years ago.

That information floored me. Now 302 out of every 100,000 Americans are choosing to end their own lives, almost 90,000 people this year. My question to him was how many were terminally ill people wanting to stop their suffering, but he didn't have that information. Apparently, a suicide is not yet subdivided into categories of "Why?". With almost .3% of the population opting out early, I saw the interview as a double-edged sword: I could help more of the terminally ill ease their fears, while at the same time an explanation of something better than what many feel in their unhappy lives might be interpreted as a permission to go there without consequence to the reason they are living.

I wrote to the reporter the following letter, which came from my processing of the experiences of those who were able to tell of their experiences as they passed on, as well as the teachings of an amalgam of spiritual leaders.

My Op Ed piece:

These are my feelings, my thoughts. Thye science speaks for itself. However, since I am constantly asked my opinion, I will share it here.

Where we go is better. It will be better than living. We are here in this plane in order to learn and grow and experiment in this thing called life, with all its pain and suffering and joy. It is something we may choose to do before we arrive and we should strive to understand while we are here. Do we take that experience with us when we leave? I don't know for certain, but I believe we do. We are here on this Earth to experience life, to embrace all that it entails and learn from our experiences, to be one with those experiences: the joy, the pain, suffering, elation, achievement, success, failure, and then death. It is our humanity. If we chose premature death as a capricious solution to a transient problem, then we are not experiencing all life has to offer or the reason we came. Your body will tell you when it's time to consider an early retirement from this experience known as life.

Dying is the most important moment that exists in any life. We should approach death with grace and respect. There is nothing to fear. We used to fear the unknown, but now we know what lies beyond,

So be present, be here in this moment. Don't anticipate what lies ahead. Don't yearn for things of the past. Let the past go, with forgiveness; and let the future go, without anticipation. What lies beyond awaits us all as pure love. Each of us contains a being that doesn't die, and a being that does die.

Everything must and will die, except the soul. Your preparation for dying is done by identifying with your soul, Not your ego.

There are realms other than this one we are meeting on. Realms where we meet the souls that we have known in the past as just souls. We won't meet them in their clothing of mother or father, or uncle, or aunt.

Then, sensual planes. Planes with color, music. Planes which have no form. This plane is where we are when all of it starts. This plane is the womb, the beginning of things. This plane is ecstatic, is the ultimate creativity.

It is important that you not be so overwhelmed by the processes of dying. embrace it with grace. It is one of the most natural things we will do.

Michael Brown

Hood River Valley, Oregon

The letter was published as a compliment to the reporter's story as an editorial opinion. The story introduced to the public the new method of self-administration and the ability of people to decide the where, when, and with whom they would end their lives. The reporter reiterated our findings and painted a warm picture of the possibility of ending one's life with dignity and at the time of their choosing.

Reaction from the powerful was swift and harsh. Organized religion doubled down on their dogma of everlasting damnation for those who commit the mortal sin of taking one's own life, and condemned my research as the work of Satan- That I was in collusion with the devil, maliciously promulgating the scenario of a happy ending for all when the Bible clearly states the truth of heaven and hell and the only way to achieve the former.

The medical establishment, anticipating a huge loss of their most profitable segment, attacked my findings as fantasy and the option of suicide as a trip back into the dark ages for medicine. They claimed that opting out was going to cause irreparable damage to the research of curing disease and prolonging life expectancy.

Then came the politicians that were controlled by the medical industry and the religious right. In a fever of haste numerous pieces of legislation were introduced in several states and in both congress and the senate to restrict or criminalize the act of euthanasia. Senator James Inhofe of Oklahoma introduced a bill that would make suicide punishable by forfeiture and seizure of all property the criminal deceased owned, and their last will and testament nullified.

What was on the federal level akin to “don’t ask, don’t tell” was being yanked out of the realm of ignored legislature and into the forefront of political debate, making the issue a divisive reactionary war similar in force to the impact of Roe vs. Wade and (later) the fire over proposition 8 and same sex marriage.

States which had never had an interest in whether someone took his own life were now having to introduce to their constituents new laws making the act a form of murder, complete with punishments to the family of the deceased. In Oregon, where assisted suicide is legal, legislation was introduced, to the cheers of a fringe minority likely bussed in from out of state, reversing the state's former decision and criminalizing not only the deceased but the family and making assisting in a suicide by a doctor cause for revocation of his license.

As much as I had faith in the common sense of the people and their seeing this sudden frenzy for what it was, I also realized that power, money, and corruption often have more sway than our citizens when it comes to what's best and for whom. So we did what we could, to do what we considered the right and honorable thing, by continuing with our work and resisting the wave of regressive politics forming against us, our work, and the suffering of terminally ill patients everywhere.

15.

As the pressure mounted from the powerful and new laws were being debated all over the country, a new bit of information came to light which stunned the world. A report had been compiled by the FDA with help from the CDC and the NIMH (National Institute of Mental Health). It stated that in the last twelve months there had been 101,820 suicides in the United States. Of those, almost 70% were via the Gateway device. The year before there had been less than 40,000 self-inflicted deaths. It was indeed a huge jump, and without doubt partially the result of our research and findings.

Although the number sounded staggering, putting it into perspective is important. Of all the people who die each year in the United States, by far the biggest killers are Heart disease, 611,000, Cancer, 600,000, Respiratory diseases, 151,000, and accidents or unintentional injuries, 130,000. I needed to justify the numbers and dove into the statistics as deeply as I could, finding that there was actually a reduction in the top three causes of death which somewhat corresponded with the rise in suicides. After taking everything into account, I saw a 30% rise in the suicide rate in those wanting to end their lives without having a terminal disease. It was not as bad as the FDA report made it seem, but it was indeed bad news.

Although I had no statistics to back it up, from what I deducted from the numbers available I believed that our research had been a catalyst for maybe 25,000 people who were too poor or in some other sort of stress-filled situation in life, who saw the place of pure love beyond this life as an escape from their torment, and the unfettered availability of the Gateway device was their ticket. It brought me to think my reading on the early Christian church, spreading their message of an exquisite paradise that awaits after death, and the slaves of the rich killing themselves in droves in order to escape their torment and realize the paradise they would never find during their tortured life on Earth.

The early church may have been telling too much truth, so those in power had the truth modified to suit their needs. All six Biblical suicides are either deemed as honorable, just, or at the least merely ambiguous. There is no condemnation of taking one's own life in either testament. It was the insertion of non-specific dogma, or tradition, which evolved the practice into a sin, then later into a civil crime. Constantine in 325 AD pushed the unified church to adopt the doctrine that suicide was a sin, and after St. Augustine wrote "City of God" in 426 AD that it gained full momentum as being a sin against God and natural law.

In contrast, scientific and philosophic works show that death is essential for the survival of the species, whatever that species is, as well as for the entire planet. From apoptosis in our cells to complex eco-systemic life cycles that feed our oxygen supply through massive plankton blooms in our oceans, death is one of the most critical parts of life. Evolution depends on it: the weaker of a species die more easily, creating a mutation of stronger life forms. We became who we are through that process; the humans more susceptible to disease died off faster than the more resistant ones, so the next generation to breed and grow were now more resistant to disease, and so it goes.

When a being feels they are too weak or old or healthy to contribute to the advance of society, the natural tendency is to let them die. In animal societies, those who cannot keep up with the pack naturally fall behind the rest and are picked off by predators and scavengers. This has always been the natural way. Primitive human societies have done the same with their sick and elderly. The Eskimos until very recently had the tradition of putting granny out on the flow ice once she was a burden to the clan, unable to be of use to the group. It was not considered a sin; it was the natural and honorable way to die of old age. Now with modern medicine we are able to live longer, and sometimes our golden years are outrageously productive, but in the case where Granny is in her 90's, being ravaged by an insidious disease that tortures her every moment she is alive, why is it now considered respectable to keep her artificially alive despite her wishes to the contrary?

There should be a graceful way to opt out when we know we are no longer healthy enough to be a contributing member of the pack. That's what my team was offering to society, the right for someone to die with dignity and grace. The formula is complex, and not for me to be an arbiter of what is right. The husband of one of our patients said that apart from the emotional devastation the disease had flooded his family with, there was the guilt his church had laid on the family once they had decided to have our procedure. But, he said, the worst part of their ordeal was the financial devastation they were suffering-their savings long gone, their home sold long ago, and the $700,000 in debt they had incurred to postpone the inevitable passing of his wife. It was a blow he would likely never recover from. By his wife's decision to have us assist in her ending her life, she had stopped the tidal wave of negative by-products of her infirmity for all of the family.

There is a difference between grieving for someone after a long illness and the person's decision to end their suffering. While here in La Paz, I was befriended by a local family of five brothers and sisters in their 50's. Their mother was 86 when I met her, grieving the recent loss of one of her children to colon cancer. As I got to know this family one of her sons became ill and died after several weeks of suffering. Her loss of two children was inconsolable to her, and one day she decided she no longer wanted to live. She shut herself in her room and refused to eat. The family tried to cheer her up and cajole her into taking food. She refused, stating clearly that she wasn't depressed, she just didn't want to hang around anymore.

While fasting is understood by the body as a natural part of existence, an 86-year-old woman does not respond to it well. While it would take at least 40 days for any harm to come to my organs, her frail frame started to disintegrate after her eighth day of fasting (this is another example of apoptosis on a grander, tribal level: The old and weak members will not survive an extended hunt for food, making the rest of the tribe less burdened by them and the food, when finally found, shared by less members, effectively converting into larger portions for the others.).

While I was prohibited from entering her room and examining her, I was right outside it and able to discuss her situation with her children. They knew that I was a doctor, but nothing more. My past was unknown to them and I needed to keep it that way, so I made no mention of my experience with death and what awaits. As delicately as I could, I introduced the option of letting her go now rather than later. They all said it was against God's will, and that they would then lose their mother. I gently discussed how granting her wishes would end a whole lot of suffering, not just for her but for the entire family. That honoring someone's wishes to end their life would be a thousand times easier than putting her through weeks of having her body shut itself down. They couldn't grasp the difference.

Seventeen days into her fast, now with her passing bloody stools and her body becoming one big bruise, I suggested that we/they assist in putting an end to her suffering. When I saw an opening I suggested they ask for medication from a veterinarian, used all the time for euthanizing pets, and asking her if she would want the injection. Their reactions ranged from moral horror to legal consequences once the autopsy revealed that she had been poisoned. I counseled them that first came her wish to end her life, so God would not judge them for doing as their mother asked. Second, there would be no autopsy of an 86-year-old woman that the doctors had been visiting regularly with the knowledge that it would be a matter of days until she expired. To their expression that suicide was a mortal sin, I responded with words similar to what I mentioned earlier in this chapter.

In the end they refused my suggestions, maybe hoping she would rescind her decision, maybe because they wanted to be with her as long as possible. When she finally expired after 30 days, there wasn't much left of her, she had gone blind, her kidneys had failed, and probably all of her other organs had either failed or were on the verge. Her blood was poisoned and escaped in her last few moments through several of her orifices. Her last week was one of constant agony, her moans and cries of pain sneaking out of her room through the space beneath her bedroom door. What was left was almost solely a bag of bones. A call was made, a doctor came and made his pronouncement, and soon the undertakers came to prepare the body for burial.

I attended the funeral. One of her sons came to me weeping and said that God was happy with the family's decision to keep her alive as long as they could. A couple of days later at the Mass, her daughter took me aside and said she regretted being so selfish, and wished they would have had the courage to honor her mother's wishes. I replied that her dying was a lesson for the family to learn. She knew what she wanted to do, and that now she is in a better place.

16.

As the political pressure in Oregon and across the nation grew, we braced for whatever consequences might arise. We copied files and hid them in a storage unit a friend of Sally had rented. We made plans for any contingency, including getting shut down or raided.

And then it happened. About six weeks after the political tornado started I had a visitor come calling. Around 9:00 in the evening I heard a knock on my front door. I cautiously asked who it was, and he answered stating his name, and that he was a law enforcement agent who had been assigned to a multi-agency task force. He showed me hid ID as I opened the door. When I asked what I could do for him, He told me that he believed that what we were doing here at the clinic was a needed service for those suffering into their demise.

He Said, "An aunt of mine was taken care of by you a couple of years ago, and my family still talks about the logic of her decision and the caring way you helped her through her passing. As a civil servant, my beliefs often need to be put aside in the line of duty, and although I don't agree with the mission of this task force, I will do my job to the best of my ability. But in order to repay the favor you did for my aunt, I've come to let you know in confidence that we will be performing a raid on your property and confiscating everything we can in order to make a case against you and your practice. There will be about 24 agents, and you, your research assistant, and your nurse will be arrested; the charges are still being compiled".

He said the raid was planned for two days hence, possibly three. Our phones were tapped but we were not yet under full surveillance. He suggested I have my lawyer on site when they come, then said goodbye and walked out into the darkness.

I sat in my living room and processed his message. All I had worked for would be taken away, I'd be faced with a legal and possibly criminal battle that could last for years and bankrupt me. My patients would be harassed and abandoned in a flurry of police activity which would surely cause them anxiety and doubt. And although I imagined that the elite power brokers were pulling strings to make this raid happen, the media would feed the people whatever story they were told, corrupting the benefits of what we had worked so hard to bring to light.

Around 10:30 I drove into town and bought two pre-paid cell phones from the all-night liquor store. When I returned, I sent a message to Max through the dark web, letting him know our plight and asking for advice. No reply came. I copied all the important phone numbers from my phone, then went to the office to start putting things in order. At 6:00 AM I drove to Eric's house and woke him up to the news of the raid and my plan to counter it. I went back to the ranch and worked inside my house until it was time to head over to the clinic.

When the crew came in I suggested a morning meeting in the garden, where I broke the news to them and told them the gist of my plan: emergency relocation to a safe and for the moment undisclosed place. We would finish the day's procedures, and call as many of the coming patients as we could on the pre-paid cell phones, referring them to a doctor in downtown Hood River to reschedule their appointments. I took 500 kits and drove to his office, explaining that I had an emergency which required my leaving town for a while. He agreed to do what he could to take on my appointments. I suggested he hire additional help or call other doctors to spread the appointments out.

By the time I had returned to the clinic to perform the afternoon procedures, Eric had arrived with the big rental truck, and with the help of a dozen day laborers from the

Depot, we loaded everything possible into the truck from the clinic and the house, and that evening made our escape into the interstate highway maze. Sally, Jenny, and Eric decided to come with me. Diane opted out, saying she could still help patients here in the Hood River Valley, and Richard stayed behind, feeling it would be best to have someone in the office to handle calls and greet the law enforcement folks when they came to do their raid.

With still at least a day in our favor before the authorities would come and find us missing, I got into the truck and started driving south on highway 35. Eric, Jenny, and sally met me in the gas station in Parkdale, a few minutes south of the ranch. There I asked them again if they were really in for a huge disruption of their lives. They were. I told them that they would need to leave their cell phones at home, pack what they could fit into their cars, and meet me at the Best Value Inn in Klamath Falls. We would meet at the diner there at 10:00 in the morning. I gave each of them $300 for the trip, explaining that they could not use their bank cards for the time being. I suggested they buy pre-paid cell phones so we could communicate during the journey, and gave them my new number.

I got to the motel in the middle of the night and took a room, sleeping until 9:45 the next morning. I got showered and dressed and went to the diner. The three of them, and Jenny's 12-year-old daughter Kayla, were at a table waiting. I greeted them with hugs, the first time we had hugged since our early experiments. We were all filled with anxiety and excitement. Being on the lam will do that to a person. I thanked them with all the gratitude I had within me for joining me on this trip. Jenny quipped that it sounded like a lot more fun than jail, and we all laughed.

Over breakfast I told them of the Mexican Government's liberal stance on medical clinics there, and if they wouldn't let us operate, at least they wouldn't arrest us and confiscate all our research, since we would not have broken any of their laws, and we could at least regroup and find a safe haven to continue our work. We would caravan down from here, and until I heard from Richard, we would take interstate 5 to San Diego and then cross the international border into Mexico. I asked them not to use their laptops and keep them completely off until we were safely south of the border.

We moved most of their belongings from their cars into the truck and headed towards California and the highway south. I called Richard around noon. He said all was well, that he and Diane were answering phone calls and emails, but there was no irregular activity. We stopped for the night at a motel a couple of hours south of Sacramento. Eight more hours of driving should get us to the border. I had crossed the border years before and remembered there was no form of passport control upon entering Mexico. I was hoping it was still the case.

My uncle Bob had suggested from my first day with him on the ranch that I keep a stash of what he called "Mad Money" around me at all times. He told me that he kept most of his savings in cash under his mattress, and every time he got his pension check he only deposited half and took the rest in cash so he could make his mattress more comfortable. His insistence got me into the habit of taking cash every month and hiding it in the orchard.

Before he died he told me to take his saved money, that it really was "under the mattress". A week before he passed he closed his bank accounts and deposited his funds into my personal account, but not before determining that the bank could only liquidate $7,000 into cash, which he handed me that night at dinner. I never counted his stash, nor mine, but now in a motel room somewhere in central California I opened the two duffel bags and counted my available cash. Uncle Bob had accumulated a $56,600-dollar mattress pad. I had been taking $3,000 a week out of the bank and burying it in a cellar in the orchard for more than three years. As I counted I was amazed at how much money could fit into that bag. I had saved $424,000 in cash since coming to the ranch. That was more than what I had in my retirement account and my bank accounts. Uncle's wisdom now came shining through my stress of being on the run. This kind of money could buy me most anything, especially in Mexico.

17.

We were in San Diego when Richard called, saying there were a bunch of cars coming up the driveway with lights flashing. We wished each other good luck just before he threw his cell phone into the Koi pond. I imagined him politely smiling and welcoming the SWAT teams onto the ranch, then asking for their search warrant. Since he and Diane had no idea of where we were, it would be easy for them to not divulge much. We agreed that they would say as little as possible. They didn't know where I was and that was all. We would be across the border before they could figure out that we had fled.

Welcome to sunny Mexico. We drove to Ensenada, on the Pacific coast of the Baja peninsula, and found it welcoming. Although nothing like the lush setting of Northern Oregon, it did have the beautiful Pacific Ocean to contrast its scrub desert landscape. Ensenada was only an hour and a half drive from San Diego, yet fully away from the reach of the American Government, who have been at times bold enough to kidnap Mexican drug lords from border towns in order to gain an edge in that particular war.

Mexico's border with the United States is host to many clinics promoting health and cures for diseases. These places exist because the United Stated government does not recognize the treatments as approved for use, usually based on the capricious recommendations of the AMA or big Pharma. So, treatments that have no profit motive for the medical establishment will not get approved. Natural treatments will not get approved. But people have faith in these alternate treatments and come by the busload every day seeking health and help. I didn't worry about what others were doing nor the efficacy of their programs, but their presence meant that opening my clinic wouldn't be too difficult, especially since we weren't providing any treatment or cure.

While the news of our disappearance made the national media, complete with sensational and titillating hints of wrongdoing and all the rest, I went to talk to the governor of Baja California. He said that 'for now' there would be no problem with us opening a clinic in Mexico, providing the President and National government had no issues with it, and of course the American government didn't put too much pressure on Mexico to shut the clinic down. He doubted they would act, though. He then mentioned another possible problem, that of the Catholic Church. Their position on life, and the premature termination of it, are well known. The people of Mexico are generally very progressive in their thinking, but the Catholic Church still rules their collective conscience.

We set about the process of opening our new clinica. I found an old house just north of downtown, right on the beach and off the trans-peninsular highway. The rent was $600 a month, so I signed a lease and we did a remodel to get the place looking more like a clinic then a vacation home. There were bedrooms for each of us and three other rooms we could use for seeing patients. Our business license was approved and we were ready to let the world know where we had relocated. My lawyer finally gave me the OK after his exhaustive discussions with the FBI and the Department of Justice. The politics of the United States, it seemed for now, would stay within the borders of the United States. Pending charges were being reviewed and he felt would soon be dropped.

However, he said, the law in Oregon was going to be strictly enforced from now on. The law stated that only residents of the state could partake in assisted euthanasia, to which we required a copy of the patient's driver's license. The great majority of our guests had their primary doctors, specialists, and their entire medical history in other states. In Oregon, a driver's license can be had if one owned a mailing address within the state, to which there were many mail forwarding businesses within her borders. For an annual fee this business would give their client a bona fide address, that of the company's office, and repackage and send the client's mail to their actual out of state address. Now the state would require the primary care doctor and specialists of each patient to have been Oregonians from before their infirmity had been discovered.

So we would not be returning to the Hood River Valley anytime soon. We opened our new clinic, and as word got out, thanks again to media stories of our move south to avoid pending changes in US law, we were back in business, implanting devices into medically qualified patients. We reestablished the rhythm of our former schedule and hired new staff to accommodate the flow of patients. We scheduled no more than 50 procedures per day, half in the morning and half in the afternoon, as we did in Oregon.

Things were going well for about six months. Although there were still many who wanted to end their lives prematurely, we were still only accepting terminally ill patients, although the argument to the contrary was a powerful one. There was so much suffering out there, and people who simply wanted to have control over the moment of their demise, but the demand for the device from truly terminally ill patients kept us fully booked with appointments. And as easy as it might have been for us to expand or even franchise our cilnica, I had no desire to further complicate my life and our delicate relationship with our host country, so the terminally ill were our only customers.

18.

One Sunday afternoon the staff and I were enjoying lunch on the patio of the clinic/home, when four rather large men came through the French doors to join us. They looked like the typical G-men portrayed in movies, style-less haircuts, sunglasses, cheap suits, and totally lacking in any kind of charm or grace. The closest one to us spoke:

“Doctor, we need you to come with us. We are not asking you, and we can assure you that as long as you do not resist, you will not be harmed”. They had big guns under their loose coats, that was obvious, and they were, although polite, dead serious.

I instantly realized that I had no choice. With their imposing presence and the energy of wanting to notch it up to aggressive intimidation, I felt that their orders might have included terminating all of us if I didn't go along. It was a scary situation indeed; we were all freaked out, so I decided not to finish my lunch. The one who was talking turned to my assistant and addressed her by her full name, saying: "please do not worry for the Doctor. He will be returned to you in two days. We would appreciate it if you were not to mention his absence to anyone. Just continue with your work as if he were in the next room."

I was driven in a convoy of two big SUVs to a military airport just south of the city. There we found a big white private jet awaiting us, engines idling and doors opening as we pulled up. Two of the men entered the jet with me, and as I sat in my designated seat, the doors closed as the engines increased their timbre. Within three minutes we were airborne, the jet making a northeastward arc through the clear Mexican sky.

That's the last I saw of the outside world. The window curtains were automatic on this bird, and they all shut simultaneously. We were in the air for five hours or so, when we landed for fuel. It felt as if dusk was upon us, which would indicate we were heading east. Thirty minutes later we were flying again. I dozed, effectively losing track of how long we were airborne, but we landed in the cold of the middle of the night.

Another set of SUVs pulled up to the plane and we departed in military efficiency by auto. The airport was small, private, and now that we had landed, was darkened. We drove for about fifteen minutes through pines on a winding, climbing road. It opened eventually onto a massive clearing and a magnificent stone mansion which looked larger than most of the hotels I'd stayed in. A butler greeted me and escorted up a huge staircase and down a hallway to an elegantly appointed bedroom. He told me that he would be my valet during my stay, and to call on him for whatever I might need. I asked for a local telephone directory and a phone, which elicited a wry smile from the gentleman. He told me to sleep, and in the morning he would serve me breakfast and lay out my clothing for the meeting I was to attend.

The room was like something out of a palace: huge overstuffed four posted canopy bed with fluffy comforters and lots of pillows, rich Persian carpet covering the ancient hardwood floor, and lots of wainscoting and crown molding over the walls and ceiling. The windows were covered in heavy velvet curtains. I pulled at one to see what was outside, but the darkness outside revealed only indistinct shadows of pine-covered hills. I opened the French door/window, surprised that it wasn't locked, and breathed in the crisp fresh air of whatever forest this was I was visiting. I had no idea of where I was, except that it was possibly in the northern United States or Canada.

I suddenly felt at peace, realizing that this situation was completely out of my control. I was in this place at great expense and for some yet unrevealed reason. If it was my money or my life they wanted, this was certainly an inefficient way of taking them. Whomever brought me here must want something specific, perhaps it was an extremely wealthy businessman or politician wanting me to explain or perform the procedure for him, or someone wanting me to disclose my formula for the transition medicine, there were many scenarios that ran through my mind, and after a bit they faded into the realm of not mattering; I'd find out why I was here in the morning, so I showered and went to bed.

My sleep was interrupted by a soft yellow glow coming indistinctly through the darkness. As I gained my senses I realized I was looking through my closed eyelids and the glow must be daylight coming through the windows. What a deep sleep it was. I blinked them open to see the butler setting up my breakfast on the balcony. A suit was hanging on a clothes horse across the room. It was tailored more for a banker than a Doctor. As I went to use the bathroom I saw it had been cleaned and restocked with fresh towels. My clothes were nowhere to be found. Gerard, the butler, told me they were being laundered, to please enjoy the breakfast, and he would be back to escort me downstairs in an hour.

He bowed out of the room as he closed the door, leaving me to sit in the cool morning sun on the balcony with my meal of excellent coffee, fresh fruit and pastries, and fresh orange juice. The view was of a hundred yards of groomed lawn ending in an endless pine forest that surrounded all the property I could see. There were no deciduous trees. Canada? Michigan? I told myself not to worry about these details and focused on what I did have control over, my breakfast and preparing myself for what would happen in an hour.

After my shower I went to the suit hanging across from the bed. There were all the clothes I could need, in my exact size and well-tailored to my frame. I was brought the newspaper (The New York Times, no clues there), and read fitfully while waiting to be summoned. The air was crisp and fresh, with only the sounds of birds and breeze to listen to.

The butler finally came in again and asked me to follow him downstairs. I entered a room with a conference table, as well as an assortment of sofas and chairs. People were sitting in the latter part of the room, about eight of them, all instantly recognizable except one, who now rose to greet me with an extended hand.

I felt honored, seeing the people gathered for this meeting. There were two ex-presidents, religious personalities from TV and the Catholic Church, and the chief of staff of the current president. Also present was a Chinese political figure I'd seen in the news before.

“Welcome, Doctor. My name is Martin Van Kleef. Welcome to My home. I trust you were treated well and wanted for nothing? I apologize for the unusual way in which you were invited to this meeting, but I needed to ensure that this entire event is kept in the utmost confidence. Please take a seat here. I'm sure you are familiar with many of the people here, so I'll forego the introductions. Let's have a talk, All right?”

After I sat down, the famous prize winning economist started his discourse.” Doctor, the world economy is an extremely fragile thing, and the difference between the world collapsing in on itself and heartily growing is the finest of lines. For example, a one percent shift in the workforce will result in huge differences at the end of the economic chain. If we have 1% more workers in the labor pool, wages are easily kept low, there is more of a tax base for the government, and demand will outbid supply, resulting in more demand for products and services, which results in economic growth and more profit for corporate and governmental interests.”

“If, on the other hand, we have a 1% smaller workforce, wages rise due to the concentration of labor, there are less people to buy products, less people to pay taxes, and economic growth slows.” He had rehearsed the speech well, and continued without seeking comment or query.

"What we are seeing now, partly because of your research, Doctor, is a dip in the workforce of almost one percent. There have been over 300,000 more deaths this year than were estimated by our actuaries, due almost exclusively to an overwhelming increase in suicides. The only sector benefiting from this surge in deaths is the funeral and burial industry. Hospitals, who count on protracted illnesses for much of their planned profit, are being deprived of their fair share because of this opt out option you have introduced into the market. People are giving clinics like yours a few thousand dollars to pass away quickly, instead of giving their life savings to health care. This may sound cynical but it is truly how the system works. The impact is hard felt in the industry already, and if it continues, health care costs will skyrocket in order to compensate for the loss in revenue."

Van Kleef spoke up now: "do you understand the implication, Doctor? The economy is suffering because of what you have discovered. Please don't misunderstand me, sir. I admire your discoveries, but there is a method that has worked for a thousand years which now is threatened with collapse."

I thought of what I was hearing: Workers making a better wage is discouraged by the powers that be because they want to dilute the labor pool and keep people poor, making less money. Suffering is eased for thousands of people, their families are not forced into bankruptcy due to lengthy treatment programs which cost all they have and more, and the corporation which rules us is upset because that ease of suffering keeps them from extracting maximum profits. I figured that these people didn't care a whit for debate on their beliefs. They were here to push their agenda on me and get me on their team. I kept quiet as the next person started to talk.

It was an Archbishop from the Vatican, who introduced himself as the nuncio of the Church, the public mouthpiece of the Pope. “The Church has a well-defined role in society, and has been the moral guide of the people of the world for almost 2,000 years. And before us, there were our predecessors, who also taught morality to the masses for thousands of years before we unified the church.”

“The church has held throughout history that good deeds are rewarded in the heavenly afterlife, while bad deeds are eternally punished in Hell. Now you come along with this pop culture theory that the good afterlife is actually available to all, that there is no Hell. This little game of yours is undermining the doctrines of good and evil, and causing damage to our great institution. There are now many of our flock questioning their faith because of you”.

“Don't think for one moment that your discovery is something new, Doctor. Our scholars and mystical orders have reported the exact same findings you have concluded for more than eight hundred years. But we are the guardians of the masses, and we decided long ago that they did not need to know about what waits beyond. They need to have faith in our descriptions of good and evil, of punishment and reward, or else there will be no morals in the world at all. Without those core beliefs, there would be chaos and immorality running rampant everywhere. If your description of a perfect afterlife for all continues to spread, the church will diminish in its world position. People will think they have free will and control of their destiny. They will lose their moral compass and anarchy will prevail. We need to keep the current belief system in place in order to maintain moral control of the masses. Without our doctrine, society would collapse into chaos. You have opened a Pandora’s box, the contents of which the people of Earth cannot be allowed to see. Without the fear of eternal damnation there would be nothing stopping society from throwing the ten commandments and the rest of God’s word into the waste bin.”

“suicide is a mortal sin, a crime against God himself, and your espousal of a ‘happily-ever-after’ is pure *anathema* to our institution and the security of our society!”

He was the hottest headed of the group as far as I could tell, smiling and sweet talking, with calming suavity, but all the while displaying his contempt for me and his sanctimonious indignation with his words, "pop culture theory" and "little game". Fine with me. After all, this man just admitted that his organization willfully manipulated data and lied to countless billions of people, knowing the truth all along. Their doctrine, based on lies and guilt, was surely causing much more pain and suffering than the truth would have done.

I thought about the Buddhists, who believe that what is, just is. That the 'enlightenment' they seek to obtain is knowledge of a universal interconnectedness of all things, pretty much like the data from our patients who transitioned to the other side. Buddhists are not threatened with the dogma of eternal punishment for anti-social deeds and acts, and yet the crime rate in Buddhist influenced countries is significantly lower than in their Muslim and Christian counterparts. Maybe letting go of the big lie isn't such a bad idea after all.

He continued: "We are also seeing a decline in revenue from parishioners, less attendance, less communions in the places where this topic is most publicly discussed. It's been a steadily growing trend in the last months, but the trend is very disturbing. Our revenue from tithing is already down almost three percent worldwide..."

"This church has had an unbroken chain of direct contact with God for over 1,600 years, and now, with these ideas you are spouting everywhere that anyone can have a meeting with God, that all they have to do is *transition,* is contrary to all we have been espousing, all we've been trying to protect the masses from for all these hundreds of years. If this frivolity is allowed to continue, the church will soon lose its hold on the people of the world."

Wow! They are afraid that the truth, in the hands of the public, will destroy their organization. I flashed on my favorite scene in the film, "The Wizard of OZ", where the booming authoritarian voice bellows: "Pay no attention to the man behind the curtain!" They were manipulating the masses through lies and guilt, fearful that the truth would expose them for who they really were. I kept silent, preferring the luxury of my thoughts, realizing that debating this man, or any of them, for that matter, would lead nowhere. I started to see where the meeting was going.

Next came one of the Presidential 'twins'. They had been doing everything together now that they were no longer in office, obviously together to show the wonders of bi-partisan cooperation to a cynical nation. Today, however, they were a phenomenal display of the power of our host and his agenda for *his* new world order. The younger of the pair addressed me. During his reign as our commander in chief, I found his southern twang earthy and endearing. Now it soured my previous sentiments and grated my nerves.

"Doctor Brown, our government needs a certain balance in order to function smoothly. We've worked more than two hundred years to get our great nation to the place it is today. If we start having lots of folks check out pre-maturely because they think there's some kind of new wave afterlife out there, or to stop their suffering, it really will upset the balance of things and could severely damage this great government of ours."

"Let's face the facts here, doc. The most likely people to want to move on are the unhappiest ones. Traditionally, the most unhappy, discontented folks in our society are the poor, right? And if they all decide one day to jump off a bridge because of this thing you discovered, we would be left with two choices for their replacement, neither of which will set well with voters. The first would to be to import a new poverty class workforce from abroad, or convert the middle class to the new working class by manipulating the banking and market situation, as well as a good dose of hyperinflation. "

"The rich will do all they can to protect their interests,

but the other 99% of the country will surely suffer, one way or another". Our Government isn't equipped to handle such a dramatic shift. We need to return things to the way they were before this suicide epidemic started. We all must exercise our patriotic duty to put an end to this phenomenon as soon as possible."

The collapse of Government? Seriously?! I recalled finding out during my research that it was, in fact, the wealthier folks who more often historically turned to suicide once they realized that money didn't give them purpose in life. Poor people tend to see their struggle and work as their reason for living, and are sustained by their hopes and hard work. So the ex-president hadn't done his homework.

The Chinese man spoke next without introducing himself: "We have a somewhat different problem in the East, doctor. With the majority of the world's population, we have a plethora of people to spare, and do see the option of a future where we could use your vivid descriptions of the next world as an incentive for a lottery system where our people would vie to control the population in the name of the State. This will inevitably result in several problems for the rest of the world. Now that we are the largest economic power on the Earth, we must carefully consider the peoples of the world as a parent would consider the problems of his children, with compassion and grace. We do not want to upset the balance of the world at this time."

"We need to consider the message we would be sending to the rest of the world about our value of the human life, as well as the contrary beliefs of those who have already spoken to you today. The other issue, once again, is economic in nature. Currently, the average salary of a Chinese worker is about one dollar for a day's work. It is a wage which provides sustenance to a billion people and is thus a fair wage."

"If we decide to adjust our population, using your discoveries as an incentive to the masses, our economic model will shift to a point where Chinese made goods and services may lead to much higher prices in the consumer market".

As he went on, I realized that he was not only pressuring me, he was flexing his muscles to our host. What I was hearing was incredible. A world more in balance would make foreign goods more expensive, which would actually give home made goods more value, and increase the wage at home, giving people more opportunities, etc., but it seemed that the opinion s of these elite few had no room for the concerns of individuals or societies, they were interested in profits and complacency.

Were we all insane? They were trying to use the people of the world as pawns, and were afraid of giving them the option of a dignified death at the time of their choosing. Some of these few wanted as much as possible, others wanting to offer a glorious death in the name of patriotism, all wanting to keep the people of the Earth as a commodity, like hogs and oil. I felt a chill of terror for our world as these people plotted and contemplated the future of human rights. My head was full and spinning, rage rising within me: I had had enough, and stood up to find my way out of there.

Our host was quick to see my intention, and quickly diverted me by clapping his hands and suggesting we take a break for lunch. My butler appeared instantly at my side and asked me to follow him to the patio for the afternoon meal. The patio was large, with several tables set for dining. I took my seat at one of the tables and wondered if there had been a change of plans. No one joined me there, the others in our meeting apparently dining elsewhere. Was this designed to psychologically alienate me? Were they to have dined here if things hadn't gotten so awkward in there? I didn't care, really. I was actually quite glad to be away from the lot of them.

I still didn't know why they were putting on this grand show for me. They obviously wanted something special of me, and I knew the rub was coming soon. Why am I here?, I wondered, then managed to laugh out loud at the irony of that statement. It was the biggest question of many spiritual people, and the one I had asked of the All-knowing being in my DMT driven astral voyage. But at this moment it certainly didn't feel like I was here "to Love".

After lunch, we were all reunited in the same room, in our same seats, and now our host took his turn as orator. "As you see, Doctor, there is a great deal at stake in the world, now that you've gone and thrown your little monkey wrench into our machine. I want you to know that I've asked these people here today to help explain the complexity of the problem, and just how far reaching the effects of what you are doing extend. I'm sure you got the message they were trying to convey to you."

"What we want from you sir, is your assistance in correcting the problem you've initiated with your discovery. We have a plan that will help restore things to a more normal cadence in this suicide and assisted death phenomenon. We recognize the compassion and sincerity you have with the truly terminally ill, and see that as an acceptable fringe alternative to mainstream care for the dying. In fact, your surprise move to Mexico has been quite propitious for our plan, facilitating the shift in credibility we are seeking in regards to your work."

"You may have heard of, or even possibly met, the neighbor of yours in Ensenada, dr. Olney, whose breakthrough discovery which halted the metastization of cancer almost changed the course of medical history. You can imagine the shock of the medical establishment, which invests billions of dollars annually into cancer research as an industry, when this unknown private doctor found a miraculous and lifesaving treatment. The business of cancer is too big to let someone like him find a cure that was financially accessible to almost everyone. He came to a meeting a little like this one a few years back, and we made a deal with him that allowed him to join the fringe of non-FDA approved treatments. It meant that he would not practice in The First world, but he was allowed to keep his license. His discovery also had the potential of radically disrupting the system, so we asked him to modify his work accordingly."

"Unfortunately for us, your situation, practicing legally for so long, gaining a reputation, then those surprise media explosions, has put us in a slightly different position. Your credibility gives you some immunity from that kind of banishment, but right now that's beside the point. Some of our team even came up with the option of eliminating you altogether, but that particular was vetoed due to the risk of your possibly becoming a martyr, and pulling untold more followers over with you; what I like to call the Pied Piper syndrome. So for now, that option is off the table, not that you'd mind either way, I suppose".

"Actually, I would very greatly have minded, thank you", I replied. They were my first words since my "hello" that morning. I guess it's a good day when the masters of the universe decide that putting a contract out on me isn't in their best interests.

He barely paused, "So, what we would like you to do, Doctor, is to come up with some evidence which shows that there are 'bad trips' as it were. We need to put the fear of death into people once again. All this "I'm having a bad day so I'm going to die" is pure rubbish. People need to be tougher than that. They need to be afraid of what might be waiting for them."

He must have seen the quizzical look on my face, so he came out with his point: We want you to put down a few people who will describe their journey as going to hell, to put it briefly."

I didn't get it. I said, "We have no data whatsoever to support that at all. In our research it's always the same in hundreds of cases: Comfort, caring, the feeling of coming home and all the rest. There simply is no bad place there."

"Of course there's not... yet. We understand it. But, for the good of society, we need to create some fear-inducing evidence that will curb the desire to escape for all these people who are so eager to *transition,* as you call it. You will need to start reporting these Hellish experiences as a normal part of your research, and we'll let the media run with it from there. We need to bring this back into balance."

"I just can't add them," I said. "That's not how science works, not to mention integrity".

"Look Doctor, I'm not asking you. I'm telling you to take this report (he took a memory card out of his pocket and handed it to me) and incorporate it into your routine work. Let it sit in your database if you wish. Take my word for it: this will work. The recordings are of actual people who died while experiencing hell, all using your methodology. Believe me, doctor, the transitions they had were truly the stuff of nightmares".

"But it's not true" ... I whispered through the shock of what I was hearing.

He ignored me. "We are asking you first, doctor, because you have by far the most credibility in this field. These results will hold the greatest impact coming from you. These cases will be released to the public, and will have the desired effect. It will be much easier coming from your clinic first. In this way the whole debate would be put to rest rather quickly".

"If you decide not to cooperate, we will still release the data through another clinic. It will eventually have the same effect. We need to reestablish that hell exists and let it soak into the collective consciousness of the masses once again. Then after denouncing your discoveries and research as corrupt we will remove all mention of you from the world, and soon this all will be forgotten".

"What do you mean, remove all mention of me?" I asked.

"The media will ignore you. You and all you have ever written will be deleted from the Internet. Your work will soon be a dim memory to society, then forgotten. That is actually pretty simple for someone like me to do. Without that reinforcement, you and your little clinic in Mexico will be relegated to the extreme fringe of medical practice".

"Take this, doctor, and decide (He handed me another memory card). We'll give you a month to release the first case study of the hell experience, then another three months to release the other five cases. If you choose not to cooperate with us, we will make sure that all media outlets on the planet gets a copy of what's on the second card there. It's something you really don't want to have happen. Your plug will have been pulled and all you will have worked for all these years will be for naught. That card is filled with dirty little secrets about you and yours. They may not be true, but only you and I will know that. The media won't care, they're more predatory than sharks when they smell blood".

"Please understand that we all greatly admire you and your work on a personal level. We are all to some extent looking forward to our own transitions into death, but first we want to take that hope back from the little people out there. What we are doing with you and your research is just business, and there will be nothing stopping it, period."

Then, as if nothing previous had been said, he clapped his hands once for emphasis, and with a huge smile and homey demeanor he said: "Thanks so much for your visit, doctor. I do hope you'll see the reasonable side of this and do the right thing. I'm sure you'll agree that for the sake of security, this meeting never took place. All of us, in fact, are in other places right now, including you, doctor. You have been recorded crossing the border in San Diego, and have been actually buying the clothes you are now wearing. You will be crossing back into Mexico around 9:00 tonight, after a nice dinner with a colleague. Have a pleasant flight back, won't you"?

And with those words he turned his back on me. Gestured to the others to stand and leave the room through the far door, and did not look back. None of them looked back. He was the last one to exit the room, and the doors closed quietly behind him.

I rose from my seat to find my butler hovering behind me, “Sir, would you like to rest before departing?” he asked.

“No thanks, I'm sure I'll have time for that during my flight.” he escorted me to the front driveway, where there were the black SUVs waiting for me. I could feel the eyes on my back from the mansion as I got into the vehicle. They thought they had me. I could feel the smug grins from behind the triple glazing and the distance between us.

I did sleep on the flight, and did an awful lot of thinking as well. Their ideas and strategies were against all I believed in. The time for the liberation of the world's people had come, if only in this small detail of truth in death, and the voice within me was crying for me to rise to the challenge and do the right thing indeed.

I was given a fist full of Pesos when we landed back at the military base in Ensenada. My escort told me to leave the base, cross the highway, and take a taxi back to our compound. And within 30 seconds of my stepping onto the tarmac, the jet had sealed its door and was turning around for takeoff, shortly to disappear into the anonymity of the heavens.

19.

Once I got back to the compound, I greeted everyone and told them we should take a walk on the beach. I changed my clothes and led the group to the path that took us to the beach. I told them to assume we were being spied on by the likes of the CIA, so keep all communication as covert as possible. Our place was probably bugged, our computers monitored, and our actions watched and listened to. I told my assistants of the flight, the extraordinary meeting, and the ultimatum.

My colleagues were shocked by the story. They unanimously wanted to fight back. I listened to their outrage for a couple of minutes and interrupted, asking the group to think about how we could counteract their plans. We stayed at it for an hour, came up with some preliminary ideas, and then called an end to the meeting. We went back to work. There were patients waiting, after all.

The next day we met again. Sally had gone into town and bought a used laptop from a pawn shop, lest the memory cards were infected. She made real time copies of the cards, and put the originals in a safe place outside the lab. We went to the beach once again and watched the videos of the forged transitions. Michael saw that the heart activity was not consistent with the video, so they probably had one person on a hallucinogenic chemical while he was controlled by another drug which caused anxiety and nightmares. Meanwhile the subject with the vital signs was probably in another room going through the real procedure. The videoed subject was filmed so that the IV bag was not in the frame of the camera, so it was likely that the death was caused by a lethal injection of some kind directly into the IV. The scenes, however, looked convincing enough to anyone who wasn't intimately familiar with the minutia of the procedure.

More startling, however, was the second card. On it was video of someone who looked like me being arrested for dealing drugs. It was a border crossing video, and the car was just like mine, even to the same license plates. The driver looked enough like me to convince most folks, and the imprinted date of the video was time stamped about four month in the future. It was fascinating and very scary. In addition, there were documents that had been altered, made to look as if my research was falsified to achieve my goals. Another video came up, showing a shadowed figure in a studio setting, giving me a message.

The Voice of Van Kleef told me about a scenario wherein the Mexican Federal Police, after hearing of my arrest at the border for working with the Sinaloa cartel to distribute drugs, stormed the compound and found much more, as well as weapons and lots of cyanide laced cocaine and heroin. My team was rounded up and two of them were killed resisting arrest, while the rest got to spend the rest of their lives in a Mexican prison, which, he reminded us, weren't as luxurious and fun as the ones in America. He said the laced drugs would label me as a terrorist intent on killing thousands of people. I would spend the rest of my life in a Federal penitentiary.

It looked as if their plan was well thought out and escape-proof. We processed what we saw and thought about the consequences of each of our options. We decided to give our opinions the next day. I worried all evening about the probable resignation of my young assistants. After all, they were all highly skilled workers who could instantly get other jobs elsewhere. It was the best for them, and if they kept quiet, they would probably be all right in their new lives. As for myself, I needed to take a proactive stand and get the truth passed on to other professionals willing to keep the reality of our discovery pure.

When we met again, I was ready for the exodus of my staff to the United States in an effort to distance themselves from me. They deserved no less, to be sure. What I got was something as far from that as I could have imagined. They had decided to stay and fight. They also believed in the truth above all else, they claimed, and after all, there was nothing to be afraid of anyway. We all knew more than anyone else of the beauty and joy of crossing over to the other side.

Eric stood in front of all of us and made his point of why they were ready to fight for this. “We have sometimes asked ourselves and each other the philosophical question: why am I here? We see the beauty of the other side, so we ask ourselves again: why are we here? And the answer seems to be: to make the world better; To care for our fellow man. To experience this thing called living. Living deeply and with integrity. Maybe we came here from there, which is why it always feels like coming home when our patients go there. Maybe we were all perfect beings who decided to go to earth to learn about struggle and imperfection and impermanence. To learn about passion and the difference between good and evil, because it seems that it isn't an issue out there.”

“If we can't stand up for what is right, we aren't growing and learning in the ways that matter. What we've discovered is bringing us closer to a better society. What we risk if we cave in to these folks is just unthinkable. How many people on this planet have been freed of their fear of death by our discoveries? We have empowered the world's people, and now these elitists want to take it away and enslave them all over again."

"This thing called living must be an important part of our spiritual growth, and we cannot do it in a corrupted way. The reason we are here is not only to learn, but to teach, to lead, and to fight that which will hide the truth from our brothers and sisters. If it's true that we retain the knowledge we gain here when we cross over, I want to make sure that I can bring with me the knowledge that I did what I could for the betterment of the world, and didn't give in to the pressures of the corrupt elite".

They decided that we should implant devices in every one of us, just in case we actually did end up in a prison. 13-year-old Kayla was sent to live with her grandparents in Portland, even though as much as I tried, I couldn't convince Jenny to go with her. We started working on the seeds of a plan which might just save our asses when the day of reckoning would come. We walked for a couple of miles that afternoon, turning over ideas and plans for the battle to come. We realized that we were to come up against the full force of the biggest institutions out there, and needed to plan everything with extreme care in case the enemy were to find out what we were up to.

I figured that we were being watched very, very closely. We developed a healthy sense of paranoia, feeling that everyone on the street was looking at us. And in Mexico, that's the normal way of life, everyone staring at each other all day long, but now we felt as if they were all working for our nemesis, whoever he might be. I got to feeling that the ones who weren't staring were the actual spies, but how could we tell. It was best to keep to ourselves and avoid contact with the outside world as much as possible.

Gone were the use of phones and internet, the mail and couriers for any type of sensitive communication. We started swimming in the ocean with our backs to the shore whenever we wanted to have a private conversation. I started thinking that the only safe place left in the world was in the villages of northern Pakistan, neighboring up to Osama bin Laden, where the US government was completely impotent, clueless, and unconcerned (I realize that they finally got Bin Laden, but only after more than a decade, and after he was already dead, so it seems to be a safe bet that his village would be a perfect place to hide).

We came to a consensus that the only way to fight this war was to go on the offensive. We needed to find a journalistic source that was beyond the reach of these people and governments, and show the world what they were planning to do to discredit us. It seemed that the European media would be much more open to publishing our story than anyone else. We decided to make lots of copies of the items that Van Kleef gave me and, along with copies of our research and a letter from me describing in detail exactly what was going on between these people and my group, and send them simultaneously to every media outlet possible.

If we were being watched, we would surely be stopped if we tried going to these sources, so we needed to become spies, of sorts, and secretly get the information out. Copies were made, after re-filtering the original documents for any possible tracking devices. We put together a concise summary of our research along with a letter I wrote detailing the concerted effort to stop me from keeping the facts factual in my research. The package was converted into memory cards for easy handling and shipping. We made five hundred copies on a computer that was free of any connection to the outside world. Eric drove four hours to Mexicali in order to find an internet cafe that might not be tapped by our enemy, and after using all the trade craft we had read about in the novels of John Le Carre', he felt that he had found a safe place to use the internet.

There, he pulled off the addresses of all the media outlets of the world, then, at another, he pulled off other stuff we felt we might need in order to make our plan work, some of it of the utmost underground nature. He also reached out to Max, who had not contacted me since our flight from Oregon, even though we still ordered our kits from his company's site.

What we needed were friends, all kinds of them in all kinds of places, who could be trusted to help us in our war against the powerful men whose aim it was to crush us if we didn't cooperate. What kind of friend could we find to help combat this kind of injustice? There had to be a network out there willing and able to run with our pack against the big boys. I hoped Max would be one of them. He had hidden himself from the world while simultaneously making deals with the world for manufacturing and marketing his devices, was running an enormous business, and wasn't affected by the law, wherever he might be.

20.

I was awakened from my sleep that night by Eric's hand shaking my shoulder. As I came fully to consciousness he whispered into my ear that he would meet me on the beach. He had just returned from his sojurn to Mexicali and needed to tell me about it urgently. I put on my slippers and walked out to the moonlit beach in my pajamas.

"I got a reply, Doc", he exclaimed in an excited whisper.

"Max got back to me within a few seconds. We chatted for an hour, and the long and short of it is that he is a part of a small and powerful worldwide network of committed, brilliant people dedicated to speaking truth to power, that can help us fight these people. He said that we would have to pre-empt their attack by exposing it all, and that he could assist with the details and more. I sent him a copy of the files. He said he would be in touch in a day or two, and that you should be ready to travel.

He also suggested we turn the clinic over, to train locals and get them up to speed, then gradually disappear".

I had Eric go over everything with me twice, until the point that I felt like I had been there with them. Max was a member of the group: "Libertas Vero", which translated from the Latin means: Truth is Freedom. Later I would search for them online. The web had scant mention of the group, but the dark web was buzzing with their activity. Libertas Vero was a community dedicated to keeping governments and corporations from distorting the facts or conveniently spinning false stories to promote their corrupt agenda. Their database was huge, encompassing what looked like every facet of misinformation, dishonesty, and corruption across the globe. It had the feel of an underground Wikipedia, and with the volume of information they were amassing, it was clear that this group was well embedded and dedicated to their cause.

Three days later came a delivery of medical supplies, our regular order for the month. Laminated within the packaging was a micro SD card. Eric knew where to look for it, or it would have never been found. That afternoon it was loaded into our laptop, revealing Max's plan for our getting the word (and ourselves) out. Max urged us to quietly but immediately divorce ourselves from the clinic in Ensenada, to leave it to our Mexican employees to operate, and for the American staff to one-by-one return home to their friends and families, re-assimilating into the lives they had before they started at the clinic. His plans made detailed suggestions for each one of them. For me he had arranged for an invitation for me to speak at the royal academy in London, giving them a report on my technical findings. He suggested I bring copies of the files we had compiled, that there would be allies there who would be helpful in disseminating the information with us. Max told us how to package the SD cards and flash drives so they wouldn't be brought to the attention of airport TSA personnel or the US Customs officers as we boarded planes and crossed the international border.

The following morning at exactly 9:00 the clinic phone rang, the caller being a very British-sounding lady, a secretary from the esteemed Royal Society of science in London, explaining that the scheduled speaker for this coming Sunday would not be able to attend, and would I be so kind as to take his place. The members had expressed great interest in my findings and were hoping I would be able to enlighten them in person. After shuffling papers around my desk for a few minutes to make it seem as if I were reviewing my schedule to see if I could spare a week away from the clinic, I accepted her offer. I would leave in two days. She asked me to forward my flight information, saying I would be met at the airport and accommodated from the moment I touched down until my return to Mexico.

Deciding to avoid crossing the international border at San Ysidro (Between Tijuana and San Diego) and American customs officers, I took a flight from Tijuana to Mexico City and on to London. In his message to us, Max insisted I purchase a first-class ticket to London, explaining that it was important to be as rested as possible when I arrived. And although $9,000.00 was an outrageous luxury for eleven hours of travel each way, I rationalized that my life might end at any moment, so I might as well go out in style. Maybe it would have been cheaper if I had bought 15 days in advance.

I was met at the airport by a chauffeur holding a tablet with my name on it. He drove me to the Savoy hotel in a gorgeous Bentley limousine, so for the moment I was thoroughly enjoying this new chapter in my life. And max was right; I felt pampered, rested and clear-headed without the usual jet lag that comes with eighteen hours and almost 6,000 miles of west to east travel. At check-in, my attendant told me that I was invited to enjoy the in-room bar, and that my laundry would be cleaned and pressed daily. The pampering continued on into my hotel room.

That night I received a message from Eric saying he had an appointment with a publisher in New York to discuss his book. It was his code for a contact willing to help with our quest. I worried for him. Perhaps he had been in contact with Max again, and that being in New York was going to guarantee that our information would be given to multiple sources simultaneously. Hopefully the Van Kleef secret army wouldn't suspect that our movements were anything other than what our cover stories explained.

Sir Brian Sedgwick, one of the directors at the royal society, had invited me to dinner to discuss the program for Sunday's event. I was to give a 30-45-minute talk on my findings, blending the data with the stories from our patients, and to try to entertain as well as inform. He told me most members weren't scientists per se, so the data should be presented in a more everyday light than I would present for peers. There would be about 15 minutes of audience Q and A, then later a cocktail mixer with the members in the ballroom, about three hours' total.

I spent Saturday sightseeing and Sunday in my hotel room preparing for my talk, all the while anxiously expecting a tap on the shoulder, knock on the door, or a phone call from some clandestine contact person. The waiting, the anticipation, was exhausting. My placid exterior hid an anxious inner me that was hyper vigilant of all that occurred within my line of sight. I could feel the watchers ghosting around me as I met with Sedgewick, wandered the city, and worked in my room. I imagined my phone and hotel room were bugged, and that my talk at the royal society would be peppered with agents from both sides of this microscopic cold war I was the center of. And through all that waiting and anticipating and anxiety, I was never contacted.

Sunday evening came at last and my stress finally dissolved as I was brought again by limousine to the front steps of the royal society of science, my mind now filled with the task at hand, my speech to the society members. Sedgwick met me at the entrance and escorted me to my waiting room after a cursory tour of the facilities.

"Full house tonight, sir. There appears to be quite a bit of interest in your presentation. The event sold out quickly and we were asked to provide more seats and standing space for the unusually high demand for tickets. We've even gotten threats of protests, so please be prepared for some incredulity and debate from some of the members this evening sir." He made it sound like a normal state of affairs for someone speaking on a controversial subject, and reminded me that without conflict and skeptical debate there would be little scientific progress.

I replied that I'd been defending our findings and

debating "God's plan" for a while now, and my facts would speak for themselves. "I'm not here to debate the existence of god or if there is a divine plan, I'm here to inform the world that we have glimpsed into the afterlife and that there is nothing to fear in dying."

"Good luck with that this evening sir, I'm sure you will make a compelling presentation." And with those words he politely left the room. I wouldn't know the difference between Edwardian and Victorian design, but it seemed to me that all sitting rooms in England had the same interior designer: Walls filled with floor-to-ceiling bookcases, portraits of distinguished gentlemen painted Long ago, Persian carpets, old-school furniture, and the obligatory lacquered mahogany desk that had to weigh 500 pounds. The two windows in this very tall room had heavy light-blocking curtains that went from ceiling to floor. There was also the cliché sidebar with an assortment of crystal decanters half-filled with amber or clear liquids, highball glasses and a filled ice bucket. It felt like a clone of every British period film I'd ever seen. I smiled at the comfort it brought.

I took a club soda and reviewed my notes. Eventually there was a gentle knock on the door. On my reply, a young man stepped inside the threshold and said, "Five minutes Doctor Brown. I'll be waiting just outside to escort you to the salon." Gathering my notes and my composure, I joined him for the short walk to where I would be speaking.

It was a large room filled with maybe 500 people. There were folks standing in the back and along the side walls as well. Sir Sedgwick went to the podium and addressed the audience to introduce me. As he spoke I had a flash of being assassinated as I spoke, and as I put it out of my head I vowed to cut down on watching action films. A deep breath and I was replacing him at the podium.

The talk was met with enthusiasm and occasional applause, and afterwards there was little in the way of dissent or disbelief. One gentleman stood up to ask a question and insisted there must have been some negative experiences, perhaps some "bad trips?" he queried. It gave me a paranoid chill to hear Van Kleef's words repeated in this public setting. I took the opportunity to hesitate before answering, as a tactic to appease those waiting for me to reveal negative inputs. And then I stated that there had never been an instance of a negative experience in the entirety of our research. The thought of a bullet meeting me once again flashed into my brain, and I felt feint for a long moment.

Sir Sedgwick then relieved me of the podium, begging applause for my efforts. The young gentleman escorted me back to the door of the sitting room, adding that he would be escorting me to the ballroom in 15 minutes so I could meet the members in a more intimate setting. He added that he very much liked the presentation and now felt more comfortable with passing, thanks to me. I thanked him and entered the room.

21.

I breathed in the comfort of this large yet cozy room. After an hour on my feet I needed to pay them some attention, so, sitting on the sofa I took off my shoes and rubbed my feet. I laughed out loud that my socks were mis-matched. It seemed that I was quickly evolving from a doctor to an eccentric professor during my brief stay in London. A walk on the carpet felt grounding, and as I paced I moved one of the weighty curtains away from the window and peered outside. The night was now misty with a light drizzle. I could see the main entrance to the building off to my right, and there were a group of perhaps eight people holding up signs along with their umbrellas. There were indeed protestors gathered to express their disagreement with our findings. Well bully for them; I would make a point of stopping to empathetically chat with them when I would come to leave the building.

A familiar gentle knock on the door and it was time to meet the public. My young escort told me that we would first enter an anteroom where I would meet some of the most esteemed members of the royal society and other VIPs, and from there we would continue on to the main ballroom to mingle with the other members. Sir Sedgwick was once again waiting at the entrance to introduce me to the distinguished members on the other side of the door. Perhaps the jet lag had finally caught up to me, or my stress of my ulterior motive had affected my brain, but at the mention of "distinguished Members", an image of a bunch of penises in bow ties planted itself into the laugh center of my mind, and I could barely control my chuckling as we entered the room.

The thought faded instantly as I was met by one of the royal brothers and his lovely wife. Sedgwick didn't prepare me for this level of introduction, so I resisted the urge to bow to him, and giggled at the thought of thinking that. There were perhaps 15 people in the salon, half of whom I could recognize by image if not by name. Next in line were a pair of actors, then members of parliament. The last person in line had a face I had seen before but could not immediately place. He was introduced as Sir _________, Lord of something or other. He was a popular author whose books I had read. Instead of a woman at his side, he was attended by an assistant (or secretary) who stood at his employer's flank and was not introduced.

Sir ________ said, "Thank you for coming on such short notice doctor Brown. I am keenly interested in your work indeed. But then I imagine that most of the population in my age group are just as fascinated in your discoveries as myself. I hope the trip you made wasn't too tiring on you, you look a bit fatigued. He extended his hand and said: "It is indeed a pleasure to meet you doctor. I hope you enjoy the rest of your stay in our country." When we shook hands I felt a piece of paper pressed into my palm. As he turned away to chat with another guest, his young assistant, who looked more like a bodyguard than a secretary, quietly said to me: "yes sir, the restroom is right over there, out the double doors and to the left", and turned to join his master in the minglings of society.

Stifling my uncertainty of his remark, I nodded and excused myself to Sir Sedgwick and went to the bathroom. What else was there to do? I thought of Van Kleef and his extremely long tentacles, and wondered if he was sending me a note through this popular writer. Once in the toilet, I opened my fist and plucked out the hand-written note. It said: "I am on your side. We need to talk. Please trust me, and go into the room across the hall from this one once you have flushed this note down the john. A friend is waiting". There was no signature or initial.

I sat on the toilet and contemplated the ramifications of the note. So what was this? Either Max or Van Kleef could have come up with this plan. None of our counter information had been released yet, so was I safe? Was this to be my end? I took a deep breath and held it for a moment, then slowly let it out with a sigh. Someone entered the bathroom and fed the urinal, then washed his hands as I sat there in the stall, pondering my next move. As the patron exited and the door hissed closed I closed my eyes and let go of my fear of the unknown. After all, all my research so far had shown that there really was nothing to fear beyond this realm of life, so why should I fear another meeting with my nemesis? He had already promised me a month to start my falsifications, and this was only fourteen days into that period. Maybe he was going to tell me he had changed his mind and I was right…

I stood up and donned my jacket, and with a grin and another dep breath left the restroom and crossed the hall as leisurely as I could muster. I turned the doorknob and entered the room. There were two people sitting this rather cozy living-room, a very professional looking, very pretty young lady, and me, relaxing on a sofa. I resisted the overpowering urge to mash my fists into my eyes and rub the image out so I could have a better look. The pair stood up. The young lady made the introduction: "Doctor, my name is Elisabeth Moore and

I am an associate of Sir_________. As you can see, this gentleman is you, and he is on the verge of coming down with a flu, which will keep him in your hotel room for the next couple of days. Please give him your room key and identification, then follow me".

I was stunned. This guy looked like my twin, though as I stared I could see little differences in the cheeks and belly. He had my exact suit on, down to the shoes and mis-matched socks. I numbly reached into my pocket and withdrew the room key card and my wallet and handed them to him. She gave me a piece of paper and asked me to read the paragraph out loud in my natural voice. He repeated my words, perfecting the nuances of my speech and accent. Then he gave me a nod and immediately left the room, closing the door carefully behind him

Elizabeth walked directly to the far corner of the room and pressed on the wall, which opened with a quiet click. She motioned for me to enter and in a moment we were descending a metal ladder, which passed through three floors before depositing us into a sub-basement. There were eight doors in this rectangular room, each with a different colored circle painted on them to distinguish it from the others, as well as a fluorescent light fixture fixed to the ceiling and the ladder passing through the hole we had just descended through. In the room was a folding card table with a grocery bag on it, and nothing more. It smelled like an old wine cellar, musty and stale, like it was never used. She nodded toward the paper bag and asked me to kindly remove *all* of my clothes and put on the change in the bag, and turned away to give me privacy. I put on the pair of jeans, tennis shoes, and pull-over shirt and asked her why I needed to change my outfit.

"We wouldn't want you to be trackable, now would we"? she replied. "You never know what kind of goodies may have been sewn into your wardrobe. Now you'll be free of any pesky bugs". She brought a key from her pocket and unlocked the door with the yellow spot. We were now walking down a long hallway of rough-hewn building bricks, which looked as if they had been there forever. At the end of the ancient passageway, about fifty yards or so, we came to another steel door. Elizabeth undid the deadbolt and escorted me to an underground parking garage, where a big black taxi was waiting for us.

The driver had the turban of a Sikh, but I had a feeling this was a disguise. His posture was way too rigid for that of a cabbie, and he was obviously in excellent physical shape. He opened the rear door and we took our seats. He started the vehicle and drove us out of the garage and onto the streets of London; one more anonymous taxi in a sea of black cabs traversing the metropolis this misty night. We made our way out of the city and headed north, or so it appeared to me. Now seemed like a good time to start a conversation.

"So this seems like a lot of trouble for Mr. van Kleef to go through. I'm sure a phone call would have worked just as well. I mean, the cloak and dagger thing was impressive the first time around, but you don't have to keep impressing me with it. I got the message already", I said.

"Doctor Brown, please trust me that these precautions are indeed necessary, and if Mr. Van Kleef and his organization find out that you are not on your way back to your hotel room right about now, he will surely do his best to crush all of us. Please understand that not every politician and dignitary in the world dances to his tune. There is dissention, and as with all plots of this size, there are spies willing to share their masters' secrets to those dissenters. My employer is very active in the defense of the truth and also the right of an individual to have a dignified death if and when they so choose. We were made aware of Van Kleef's plan and his meeting with you in Maine, and we had an idea that from your reactions there you weren't going to play along. If that's the case, then you will need help, and that's what my employer wishes to give you".

She made no mention of Max. I decided to do the same. I tried to come up with a pertinent question, but fell mute. I was suddenly weary and felt the need for a bit of rest. The car continued on for the better part of an hour and finally found its way onto narrower and narrower roads until it turned into the driveway of a modest estate. The air was clean and smelled lightly of wet grass. The place bathed in soft moonlight as the driver opened the taxi door and let his passengers out. We walked into the large home and Elizabeth asked me if I needed to rest or use the facilities, then offered me a seat in the salon.

A butler appeared with a tray of tea service, pouring for Elizabeth and myself. The beverage was soothing, and as I drank I realized that I had been in a state of agitated shock, running on faith and auto-pilot. Now the tea became a brief meditation. I measured my breathing and savored the warmth and mild flavor with all my senses. My hostess was silently enjoying her tea as well, probably preparing for our next round of Q and A.

In a few moments the door opened again, and a distinguished and obviously well-bred gentleman joined us. "Good evening doctor Brown. My name is Jonathan Pengarden, and I appreciate your coming to visit me in this most unusual way. I want you to know that you are among allies here, and that we will do all we can to ensure that your truth, your research and amazing findings, will remain free of corruption and available to all people throughout the world." As he spoke the butler reentered the room and whispered to my host, who excused himself to "take a call". Elisabeth and I sat silently sipping our tea. When our host returned his face and posture betrayed bad news.

"Doctor Brown, I have just received some disturbing information". He slowly sat down in his chair and processed what he was about to say for a long minute. When he spoke his voice had softened. The prepared discourse was to be replaced with this new information. "There is no easy way to say this, but I fear the game has just changed. I don't know how to relate this news with appropriate decorum, so forgive me for any bluntness. About 30 minutes ago the man taking your place, your double, the man you met just before coming here, was, um, shot and killed as he left the royal society of Science. The killer was one of the people in a group of protesters marching at the entrance of the building. When the protesters blocked his access to the car, one of them shot him four times in quick succession, and as the group scattered, fired once more into the body. A car pulled up and the killer got in. after a brief chase by the limo driver, thcy were eventually lost. The police are investigating your death doctor Brown. As saddened as I am that one of our employees has perished, I truly hope his death will give us the time we need to protect you and disseminate your information and evidence of van Kleef's attempts to corrupt your findings."

We sat silently for a long while. He finally cleared his throat and suggested we all get some rest. He suggested we resume our chat in the morning, then got up, obviously still processing the shocking news, and bid us good night. The butler entered and asked me to follow him to my quarters. We went upstairs where he familiarized me with the room before retiring. My night was spent without sleep, taking in all the information and running it through my mind a thousand times. In the morning I realized that not only would I be no longer able to communicate with my team, friends, or family, and likely be constantly on the run until my death, but that exposing Van Kleef and keeping my research unadulterated would be my new life's quest. What other option was there?

22.

In the morning we met over breakfast. The news of my death was the media topic du jour. When the talking head on the television said the members of my team were not reachable for comment I became gravely concerned for their safety. I had left my phone at the hotel, so my only means of communication was via email. My host said using the computers where we were was too great a risk. Instead, he asked me to write their email addresses down so he could have an assistant contact them safely.

Our host asked me to walk with him in the garden. There, away from Elizabeth and the staff of the house, he brought up the reason I had come to England. "Doctor Brown, A mutual associate of ours mentioned you may have information in need of distribution. Have you heard of 'Libertas Vero'? It is a group dedicated to keeping truth alive, and we understand what you are facing-the reworking of your research, the implanting of false, fabricated results to destroy your reputation and your credibility. What you have shown to the world is too valuable to be tarnished. Now that you are dead we must act immediately. I have couriers coming later this morning to package your information and deliver it to every major news outlet in Europe and Asia. We want all packages to be delivered tomorrow morning, so time is of the essence. Your associate in New York should be meeting with our agent this morning to hand over his copies for distribution. We will let him know you are still in the realm of the living, and quite safe. Do you have the files with you, or are they stowed safely elsewhere?"

I handed the electronic copies of all we had on Van Kleef and my research along with the email addresses of my colleagues. He told me he had a package waiting for me in the house, From our mutual acquaintance. We went inside to his study to find three men affixing mailing labels to FedEx envelopes. Our host handed the bundle of SD cards and flash drives to his helpers, instructing them to get the packages compiled quickly.

Pengarden took me into another room, the library, where he opened the desk drawer and handed me an envelope. He said he wanted me to be out of the country when the information hit the media. And when the time was right, his group would resurrect me to confirm all the facts and hopefully knock those like Van Kleef down.

He made a call. While he was occupied I opened the envelope. Within was a British passport and an old cell phone, but no note or other explanation. As I unflipped the phone, I saw there was one unheard voicemail waiting for me. As I put the phone to my ear I heard a familiar voice broadcasting his message. It was Max.

"Hi doc. Thanks for going to England. Our network there will take care of everything, but we think it's best if you are elsewhere when the shit hits the fan. The people who want you to do their bidding will retaliate immediately, and could have you jailed in some foreign shithole or worse. It's best if you weren't around for a few days, at least until the dust has settled a little. Good luck. We'll be in touch soon." He had obviously made the recording more than 14 hours ago.

My host hung up his phone and told me I would be escorted to a safe house on the continent that afternoon. I suddenly felt exhausted and excused myself to rest. I awoke to a setting sun and went downstairs. Sitting at the kitchen table was an elderly gentleman wearing a tweed jacket with elbow patches, looking quite the part of a country sportsman. He sipped his tea and introduced himself as Archie Moore. Which he described as his traveling name. He said we were going on a little trip to visit my aunt-his wife in Aix, the south of France. He nodded to a small roller suitcase in the corner of the room. “I see you’re all ready to go. Shall we then?” I followed as he trundled out the kitchen door and around the house to the driveway, where we got into his ancient Jaguar sedan. There were no other cars in the driveway, no activity within the house. Reading my thoughts, he said that everyone had gone out to run errands, and that they had expressed their wishes for a safe trip for us.

We drove for about three hours, passing east of London and heading towards Kent. At a town called Folkstone, Archie pulled the car into a line of other cars nudging toward the Eurotunnel, an underwater tunnel connecting England to France. During our drive he chatted casually about me and my aunt, how long they had been married, and of my job at a London architecture firm. At the train we drove the car into one of the train’s transport carriages and sat there as the train started to move. A half hour later we were in France. My uncle Archie spoke to the customs officer in French and we were soon on our way, driving toward Bruges.

Three hours later we arrived in Brussels. It was 4:00 in the morning when Archie parked the car on a quiet street lined with tall row houses. He escorted me to one of them, unlocked the door, and showed me around the place. He said this house was safe, that only a handful of people know of its ownership or use. He said I could relax here, there was a grocer a block behind, and a park nearby. As he made his way to the front door, he turned and said he had one more thing to share with me: Eric didn't show for his meeting in New York. His whereabouts are not yet known. His belongings are still in his hotel room though, so he is likely still in the city.

And then I was left alone in this unknown house in a city I did not know. The next afternoon I woke and watched the news. There was a report of an unknown group vowing to destroy me if I didn't expose negative results of our research. Other reports suggested that we had withheld these results to favorably distort our findings, while another claimed the distortions were planted from an outside source. It was all ambiguous at the end of the day. The report of my being setup by Van Kleef was brushed aside as too incredible. There was no exposing of the political elites who wanted to keep people from deciding when they should be able to die, there was, however, in the midst of all the controversy over what was real, enough disinformation to create a reasonable doubt in the minds of many as to what lies beyond.

My new phone buzzed. It was a text saying that my friend had not surfaced. Unknown return number. I went for a walk and saw an internet café. Resisting the urge to reach out to my team, I returned to the house to ponder my plight. I was dead to the world, my research had been smudged with doubt, and my clinics were not currently operating. My goal all those years ago was to help remove the fear of dying for those who were near their end. I had discovered way more than I imagined possible during my research, and my findings became twisted, used as an easy way out for some. I could understand why the section of society wanting to check out before rising to the challenges of life should be deterred from such an impulsive decision, but to throw fear once again into the equation of a dying person's final weeks seemed to be an unusually cruel injustice.

I would need to come out and face the enemy once again. This time in public so I could expressly demonstrate that there is nothing to fear, that those trying to cloud the truth have an economic and religious agenda to keep people suffering and in fear of dying. I would need to somehow compromise, perhaps by urging young people not to depart this life frivolously, but the message needed to remain clear and true: what lies beyond is a comforting thing. I couldn't say if it would be better than living, but it should be viewed as a natural progression of out spirit, and to take that next step with grace and love for what awaits beyond.

23.

The next morning I returned to the local internet café and disguised myself and my location by plunging into the dark web and rerouting my IP address all over the world. I made new email accounts and started writing, first to my staff, then to friends in Ensenada. I also wrote to Max. A neighbor in Mexico said the clinic had caught fire in the night and burned down, no hypothesis as to what happened, but fortunately the place was empty. He hadn't seen any of my staff there for a few days.

Jenny replied later: "Hi uncle. It is SO wonderful to hear from you. Were on vacation in the Utah desert. More later. Have fun at the wedding." I guessed it was her code for being very afraid yet glad I was alive.

Nothing from Max. No calls nor texts from the folks at Veritas either. I returned to the safe house and pondered my situation: alone in Brussels with the world thinking I had met my death, my team scattered and possibly harmed, and now my future in the hands of a group I knew nothing about, with guest appearances from a former employee that had betrayed me and corrupted our cause. It was a grim state, but it was where I was.

I saw two priorities emerging from this current situation: I needed to get the truth out to the world about the pressure put on me and the reality of my findings, and I had to stay alive long enough to make that happen. For now I felt I was safe, even though I didn't know who was protecting me.

My safe house sanctuary was a row house in the heart of downtown, four stories tall and surrounded by similar gracefully old structures. On my fifth night there, about 11:00 the phone my protectors had given me buzzed, alerting me to the delivery of a text message. I got out of bed and went into the kitchen to read the note. It said that an unidentified body had been found in New York that their sources say has identifying characteristics resembling those of Eric. There was an 80% probability of a match.

I felt more alone and vulnerable than ever, and not being able to return to sleep, my mind bombarded with thousands of thoughts, I got out of bed, deciding finally to escape to the roof. Pulling on my jeans and a light jacket, wearing my slippers, I sat in the dark listening to the ambient noise of the city and contemplated my lost co-workers. The girls are gone, Eric gone, for now at least, the duplicate me killed in London, the threats of van Kleef now coming true through his control of the media, and now me in the middle of Europe alone and safe, but for how long. What if the identity of my twin was discovered?

As the fresh cool air mitigated my thoughts, I eventually dozed off. I abruptly awoke to a roaring sound and the disruption of my equilibrium. I snapped instantly awake to find myself falling over onto my side as the roof of the building warped beneath me. The roar became the reverberation of an explosion and my eyes opened to bright fire light reflecting off the buildings across the street and all around me. My safe house had exploded! I had been thrown upwards at the bursting of the building's structure, and while on my back trying to assess what had just happened I saw the sky in front of the building turn bright yellow with flames as a huge ball of orange burning gas rises into the night above the front of the building. I had been knocked out of my slippers. Within a minute I had regained my composure and instinctively retreated onto a neighbor's roof.

From there I made my way across the rooftops as far around the block as possible, climbing up or jumping down onto each successive one. Once I had rounded the corner onto the cross street, I started trying roof access doors and peaking off the front and back ledges for a fire escape. There were none. Eventually I came upon an unlocked door and, as quietly as I could, made my way downstairs. As I passed a bedroom on the second floor I heard a couple quietly talking in French. There was no light escaping from under the door, so I guessed that the blast had awakened them and they were now trying to get back to sleep. I tiptoed past and down to the foyer, where I found galoshes and umbrellas and an overcoat, which I borrowed before discreetly leaving the building.

After a three hour walk I found myself in the Schaerbeek district, comfortably away from the city center and the view of the city's security cameras. Finally, all the action movies I had seen were paying off. Although I didn't know the city, I could hear the rumbling of trains in a nearby switch yard, and thought of seeking a train station to put more distance between me and those hunting me. It would be a matter of hours before they discovered no corpse in the burned-out building.

By the grace of my good fortune, I had escaped certain death and had my wallet safely in the jeans I had on. I walked until daylight and settled into a café for breakfast, trying to assess the situation and consider my options. I had no laptop, no phone, no passport, no clothes. I counted 755 Euros and my British driver's license, along with some business cards and other notes I had been collecting over the past few weeks. I had the number of Mister Pengarden, but now figured that if I was a target of the magnitude where they would kill me twice in two different countries, they certainly had the ability to eavesdrop on whatever communication he may be receiving. For now I was on my own.

The café had no television, so I couldn't find anything out about the building which had exploded beneath my feet four hours earlier. The newspaper would have been published before the fire, so I didn't bother searching for information there. At ten o'clock I made my way to a shopping center in East Brussels and bought a new outfit, shoes, sunglasses, and backpack. I filled the pack with snacks and bottled water, as well as spare socks and underwear.

Noon found me at a cyber café, connected to the web through my VPN, where I saw the news of the not-so-safe house explosion. The front of the building was mostly gone and charred black. A gas leak in the basement which had filled the house from the bottom up until it met with the oven pilot light, causing a devastating explosion and fire which destroyed the home. No bodies had been found, but the search was not complete as of the time of the report. It was billed as an accident. Of course it was.

I stared at the screen, wanting to reach out and let those who knew me, those who cared, know that I was safe. I was stymied by the thought of my pursuers knowing that their attempt on my life had once again failed. But since all my friends but one had no idea I was in that house, they were likely still under the impression that I had died in London. I had nobody I could contact.

Then I remembered Max. I knew he was in hiding too. He had sent me the fake passport and the warning back in Oregon so he must be on my side, I thought. In my desperation to connect with *someone* I once again sent an email to his deep net site.

I wrote: MPL, I'm still standing, despite two attempts in Europe, the latest in the EU capitol this morning. I lost everything in the fire and am without direction, worried, and quite alone. If I am eliminated our work will be invalidated, as you have surely seen. I need the chance to bring the truth out again and show that the claims of my passing, as well as the forged negative results, have been greatly exaggerated. I am looking for help. Are you available?

I pressed the send button and shuddered with the fear that I may have just made my location known. It was in my best interest to get as far away from the café now, and as I went to log out of my VPN, his reply popped onto the screen.

"HELLO DRB. I WILL HELP. WHERE ARE YOU? IT IS ME. PROOF: THE NAME OF THE GIRL AT THE FRUIT STAND IS ELIZABETH AND HER SISTER IS ISABEL. LET ME KNOW. MPF

I was in shock. I let the message sink in. Back in the Hood River Valley there was a fruit stand between our ranch on Davis drive and highway 35, called Cody Orchards' farm stand. At this stand there were two young sisters who ran the store, as fresh, vibrant, and beautiful as the peaches they sold. In his first two weeks working with me, he always came to the office with a bag of fruit to share. The third week he came to work without fruit, so I asked him where my breakfast was, and he replied that he was trying to get to know the girl who ran the store at Cody Orchards, Elizabeth. She had a sister named Isabel, because their mother had been fascinated by the royal family of England. He said he finally made his move and asked her out, and she politely rebuffed him with the 'I'm seeing someone' excuse, so it was time for us to pick our own fruit from your trees, Doc. That was the only time he had ever mentioned the sisters, long before we were joined by the rest of the staff. It was really Max I was chatting with. I sighed audibly with relief.

I replied that I was in the Schaerbeek district of Brussels, and on foot after the safe house I was in had been bombed. I pressed send. Five minutes passed and he replied:

GO TO THE TRAIN HOSTEL SCHAERBEEK. YOU DON'T HAVE TO STAY THERE BUT I WILL SEND YOU A PACKAGE WHICH SHOULD ARRIVE IN THREE DAYS. THAT WILL GET YOU WHERE YOU WANT TO GO. CONTACT ME WHEN IT ARRIVES. GOOD LUCK.

The receptionist had been expecting me, addressing me as Mr. Lazarus, (Max's last name). He said the hostel was full but they had a flat available three blocks away for 95 Euro, so I booked there. He told me that he would alert me as soon as my package had arrived. I nodded my assent to each of his gentle mandates. He presented a young French lady as Andrea, saying she would escort me to the flat, which she did. I settled in, and after a shower slept until dark. Concerned that I may have been followed or compromised, I spent the night in the adjacent park watching my apartment from the shadows. Using the tradecraft I learned from reading spy novels, I saw that no one had entered the flat nor crossed the threshold. I slept during the day and shadowed the place at night.

On my fourth morning there I was awakened by a

soft knock on the front door. Andrea was there with a package, my package. I thanked her for it and once she had gone, opened the box. Here was my help, from Max's point of view. I placed the contents on the kitchen table: Two passports, one British, one American, both with my photo and the same fake name, a wide mesh belt with integral pouches designed to carry money under one's clothes, five envelopes filled with money, four with dollars and one with Euros. I didn't count it all, but it appeared to be divided into ten thousand dollar packets, for a total of $50,000. And then there was the letter. Hand written on toilet paper. I laughed out loud at the implied disposal of his note. His words were sobering.

"Dr. B, this should get you started. Everything is clean and untraceable. I suggest you change your looks as much as possible; color your hair and wear padding to give you a gut. Buy a hat with a big brim and sunglasses, you get the idea. When talking on the phone, put cotton or marbles in your mouth to disrupt your voice patterns. That is important.

"Trust no one. Those who once wanted to help you have probably been bought off or threatened, or worse. I have found a home in the depths of the web, and am in a community of dissidents, hackers, and a huge group of folks who are digitally fighting for our freedom. They know the profundity of the government's illegal interest in her citizens. The government is run by the elite, and they want total control, so they have government agencies using all the resources at their disposal to spy on people, to control people. The elite are spoiled brats used to getting their way, and our discoveries have upset them. They have given the order to remove you and me as if we were insects. They won't stop until our bodies are displayed to them or we have not been heard from for a long, long time. Their sense of control and impunity is boundless.

"There may well be nowhere left to go, no sanctuary for who you used to be. You should look to the future as someone new. If you want to keep doing what you did before, then go deep underground. But you have made the truth known and that truth has gained momentum. There are already many, many others who have taken up your cause and are helping people to die with dignity and freedom. Millions of people have bought the gateway and are in control of their own end. Let go of your past and create a new life as a retired salesman and go live on some tropical island, away from where you might be recognized. Send me your banking information when you get settled and I will safely get you your money. You have made your point. Everyone in the world knows of your work. There are many others who are carrying the torch you lit. It's time to accept the impact you have made and disappear. The world thinks you're dead. Use that to your advantage.

Good luck,

M

I read the note twice and flushed it down the toilet. Stuffing the money and the US passport into the belt, I wrapped it around my waist and pulled my shirt out to cover the added bulk. I exited through the back door and into the park, deciding to make my way out of Belgium as quickly and discretely as possible. I returned to the cyber café and sent Max a one-word note of thanks. I reread my obituary in the Guardian online. There had been no updates in the two weeks since my stand-in had been killed. I was still officially dead, but I knew that there were hunters out there searching for me.

After a few blocks of walking I came to a line of taxis at a stand in front of an auditorium. As I entered the third or fourth cab, the drivers at the front of the line yelled and gestured at the injustice of my back loading them. Tough shit. I told the driver to go, just go, which he did.

“American?”, he asked in his almost indecipherable English. “No, British!”, I replied, not worrying about trying to fake an accent. He was very dark and lean, about 50 years old, and definitely not a Belgian citizen. My guess was North African or Possibly Mid-Eastern. Asking “where go?”, I replied, “I don’t know, How about France?”.

He said something in a French so broken that I was at a complete loss as to what he was trying to communicate. He was not deterred by the vacuous look on my face, and kept chatting away alternately with me and his telephone, while driving us into the bowels of the city. He parked in front of a tenement in the middle of a ghetto, turned to me and made the international signal to wait or stay where I was. He got out of the car and crossed the street, entering a shabby apartment building. I quietly left the car and retreated to a doorway down the street, trying to make myself invisible while keeping an eye on the car. Within a minute he returned to the car, this time with a young lady. Noticing I wasn’t waiting as instructed, he looked around and called out: “Mister, mister!”

I left the protection of the doorway, and he sent the young lady to escort me back to the taxi. “Come with me please sir.”, she said. “My father wants me to translate. My name is Anna.” We got into the cab and he drove on.

She continued: “My father says he thinks you want to go to France. Is that correct?”. I nodded. “I want to cross the border quietly and hope not to show my papers, if you know what I mean.”

She translated for the driver, which provoked an animated conversation between the two of them. Back and forth they went in some Arabic tongue, his hands flailing around in rhythm with his words, while I sat mute in the back seat trying to pick up a hint of their intent. Eventually she turned back to me. “He will take you safely for 700 euros. Do you have that much?”

I pulled $1,000 from my wallet and passed it to him. He grinned and exclaimed a praise to Allah, then launched into another discussion with his daughter. This one lasting 20 minutes or so. I was tired of trying to understand and retreated into my own thoughts and anxieties. He suddenly grew quiet as they turned to me.

“If you wish my father can help get you farther than France. There is a boat that brings refugees from Morocco and Tunisia. His cousin is the captain, and my father can get you passage to North Africa if you wish sir. We do not wish to offend, but your manner and actions betray your fears and desires. Please know that we too are not legal residents of the European Union, and we must live on the edge of society and the law, a place my father knows well. If you go to North Africa, you will be able to travel elsewhere freely. You can discreetly return to Brittan or go to the United States with little effort. My uncle has many friends there."

I suggested we eat something. I needed time to process their offer. And although Max and my own instincts shouted into my brain to trust nobody, the odds I put on this family being an active part of the organization out to see my demise were safely approaching nil. I relaxed a bit, realizing at that moment that I really had no fear of death, and that the motivation for my preservation was at that moment my instinctual desire to survive, and to beat these forces hoping to silence me.

We stopped at a pub in the village of Chimay and shared some dinner. One of the perks of being on the run in Belgium is that everywhere in that country there is great beer, and the local brew was outstanding. Conversely, in my time in that part of the world I never saw Brussels sprouts on any menu.

Pondering their offer, I realized that the plan they presented sounded custom made for my needs. If they were part of the network out to get me, I'd already have been dead. I trusted in my intuition and asked about the details of his plan. Another discussion and she gave me the condensed version of his idea.

"We will take you to a town in the south of France. My father will contact my uncle and set up your transportation from there to the African coast, to the city of Bejaia in Algeria. My uncle has a boat and brings refugees over every few weeks. He will be glad to bring you back with him after his delivery in France. There may be a wait for the boat, but you will not be noticed. We will give you all the instructions once we get in touch with my uncle.

They drove through the night. I slept on and off through the trip. And by mid-morning we were in Marcelles. At breakfast, Anna wrote out instructions on a postcard, along with her uncle's French and African phone numbers. She explained that he did not speak English, so I would need to be in the correct place at the correct time or he would not look for me. To make sure I understood, we drove out to the cove where the boat would land and pick me up after discharging its passengers. It was rocky and somewhat protected, just east of point Figuerolles, with a small parking lot on the bluff above and a trail leading down to the pebble beach. He would pick me up on that beach at exactly 2:00 am, fourteen days hence.

Anna listened as her father talked on the phone with his relative. The meeting was reconfirmed and a price was set. They drove me back into town and wished me luck and the blessings of Allah before we parted company. They dropped me off near downtown as they started following the signs to the freeway northward toward Brussels.

I wandered the town and found I quiet little hotel a few blocks from the beach in the district of L'Estaque, in the northwest corner of the city. It was less than 10 kilometers from the secluded cove. With two weeks to wait, I decided to make the most of the time and become a tourist. Mornings were spent running and enjoying coffee and pastries at local café's. afternoons at the beach or a hotel pool along the strip. I kept to myself and did my best to become invisible.

A few days into this vacation I found a classified ad in the local English language newspaper that caught my attention immediately. An American couple was looking for crew to help them sail Eastward and through the Panama Canal. They had a 50' ketch and were hoping to find someone with sailing and cooking experience, preferably English speaking, who could share in the cruising expenses. I went to a pharmacy and called their local number. We set up a meeting for that afternoon. They seemed as enthusiastic at having me as crew as I felt about disappearing over the horizon on a yacht.

Zack and Nikki were what I had always imagined as the typical California couple. They were in their early

40's, fit and good looking, with infectious smiles and personalities. They had worked in the tech industry for one of the big Silicone Valley players and decided to take a year off to reconnect with their values. They were, as one might imagine, vegan. This trip was to be their last big adventure before bringing a child into their lives. I asked if they weren't too old to have a baby. They were going to adopt a 12 or 13-year old when they returned to San Francisco. Yup, a real California couple.

When it was my turn, I told them that I was a New York doctor who had lost his soul mate to cancer. Her death had left me questioning my career choices, so I was traveling aimlessly until I was able to find my way again. I told them I would be happy to volunteer for all manner of duties and had no problem sharing in the costs of the voyage. They were comfortable with my limited sailing abilities. I had sailed enough as a teen that I could handle a sheet (a rope that trims a sail), stand a watch, and steer the boat. The rest they said they would be happy to teach me. They were set to leave France in a couple of weeks, as they were waiting on a part for the boat's diesel motor. Once it was installed and tested we would weigh anchor and head out. We shook hands. I told them I was available to help on the boat as needed. Nikki showed me my cabin, saying I could move in whenever I wanted. I had paid a week at the hotel and wanted to ease into my new job as crew, so I told them I'd move aboard later.

The next day I showed up at the marina after my morning run, and spent the day sanding the varnish in the cockpit in preparation for new coats. They made lunch, which tasted better than anything I had eaten in months or longer. It was a huge salad with quinoa, wasabi peas, cranberries, and kale, dressed in a light vinaigrette. Nikki told me there was no sugar in the salad, which I might have trouble getting used to. They told me about scientific research proving sugar was unhealthy and addictive, so they had eliminated it from their diet. I replied that whatever they wanted to serve was fine with me, especially if it tasted that good.

The night of my rendezvous with the African captain came. In my curiosity, I took a cab to the landing cove the afternoon of our appointment. I hiked the hills around that beach and at sunset found a small cave in the wall of the cove's bluff to wait and observe. I pulled a blanket and binoculars from my backpack and slept. At 1:00am I awoke and scanned the area with the binoculars. Nothing unusual to be seen. A half hour later I saw the boat coming in from the sea. The note of the motor was distinct and the rumbling of it grew louder as the vessel approached. It was a fishing boat, about 80' long, heading right into the little cove as planned.

The motor slowed as the boat neared the shore. I could hear voices now, and the sound of the hull coming to rest on the gravel bottom of the bay. The captain barked orders and the crew started repeating his commands to the passengers. The immigrants jumped from the gunwales of the boat into the waist-deep water and waded ashore. The scene was briefly illuminated from the parking lot behind me. I repositioned myself and saw several vans and small buses waiting. One of them had flashed its lights to signal the illegals as to their destination. There looked to be between 50-60 disembarking passengers, all of them now marching up the path from the beach to the parking lot. As the vehicles filled they drove away, one by one, bringing hope to impoverished foreigners and a social and financial burden to the European Union.

Within 20 minutes the vehicles had all departed with their human cargo. The motor of the boat still idled as she waited just off the beach. The captain called out “Mister, Mister” several times. He shined a powerful flashlight around the cove and up onto the area around the parking lot. Finally, at 2:30, He gave one last shout for his return passenger, then turned on the running lights and put the boat into reverse. The engine revved and the water at the stern of the boat churned as she dislodged herself from the bottom and backed out of the bay. A hundred yards out, she shifted into forward and turned away toward the coast of Africa. There were no teams of assassins waiting for me, it seemed like there was nothing suspicious about the boat and her crew. I felt safer for having witnessed that option working itself out. It seemed unimaginable that if the information had gotten back to whoever was hunting me, they would have preferred to take a boar full of refugees from North Africa rather than looking for me on the beach at this secluded French cove, and there were no suspicious types hiding in the shadows this night.

Is it paranoia if they are really after you?

24.

I settled into my new quarters on the sailboat, the part arrived, the motor was reassembled, and we were ready to go. We made two days of trips to the markets and stores to top off the boat's larder. What could be frozen was, what could be crammed into the fridge, ditto, and the rest of the galley and the spare cabin was overflowing with produce and packages. We were ready to set sail. Our first destination was the island of Mallorca, off the eastern coast of Spain, for a couple of days to explore and check out the boats systems.

The trip was uneventful and soothing for my soul. I finally felt completely safe and invisible to the world. Palma was a lovely ancient city, but I was once again anxious, and ready to be out of the Mediterranean Sea. We made another stop in Marbella, on the southern coast of Spain, and after another three days and more trips to grocery stores, we set sail for the Atlantic Ocean and our next port of call, the Canary Islands. I was finally starting to feel free of the feeling of being pursued. That intense internal stress took weeks to finally work its way out of my conscious alert mind. There had been no news of me since the explosion in Brussels, not even in the deep web. My sailing companions had planned for a trip of four months between the Mediterranean and the Panama Canal, which they repeatedly insisted was flexible.

The days melted into weeks as we sailed the Atlantic in anticipation of our landfall in the Windward Islands of the Caribbean. From there, while sailing over thousands of miles of ocean, I was able to transform myself into someone stronger and more relaxed. The trip had served me to fulfill my need for adventure and travel. I no longer worried full time about being hunted, though occasionally I would feel as if I was being watched. Maybe that happens with everyone, but my experiences had heightened my intuitive sensitivity to my surroundings.

I was tempted often to jump ship when near South America, disappearing into the Amazon jungle, but now that I was near enough, I knew I could return at will. And the lure of a passage through the Panama Canal outweighed my other urges. My hosts were delightful and amazing traveling companions, and being with them was a pleasure. So, we crossed into the Pacific Ocean together.

In our months together we discussed virtually everything over meals and maintenance and standing watches. Eventually the topic of suicide and choice came up. And my input as a doctor was an important ingredient in that conversation. I considered their views as a norm of the educated majority, and was intrigued with their point of view, as they were of mine. Nikki was full of statistics. She started with the fact that there were 80 million more people being born per year in the world than were dying, but now the difference is less. This year that trend has slowed to 68 million, mostly due to the ability of people to choose when they want to die. In the past 12 months there have been record numbers of suicides, mostly by the elderly. I asked if there were actually 12 million more suicides than before. She replied that she didn't know the exact number, but it must be significant.

Zack jumped in saying that the health maintenance industry has had to rethink its financial model, because the revenue stream from the terminally ill has dwindled to a trickle of its former strength. And although they won't assist in suicide, there are clinics everywhere and legislation in many states to change the prohibition of suicide for the terminally ill. Then he looked directly into my eyes and said: "There is even a readily available device that people buy over the internet which they can implant themselves or with the help of a nurse, and choose their time to let go of life. The back-alley abortion clinics of the last century have turned into the back-alley implant clinics. There is someone in every city ready to help people stick that device into their body."

I mumbled that I hadn't kept up with the news much lately, and they continued on. Zack said that big pharma and big medicine were pissed off and doing all they can to discredit the reports. I asked which reports he was referring to, playing dumb. He laughed and cocked his head.

"You can't seriously say you haven't heard of Dr. Michael Brown and his discovery of what lies beyond? He was everywhere in the last couple of years, that is, until he was killed by a lunatic in London. I really think he was silenced because he stepped on too many toes." He turned to his wife, "Remember all the 'coincidental' news reports that came out right after his death? 'There is a Hell', and all that. It reeked of the church trying to regain control of their followers. Within a month we were reading reports of huge increases in suicide rates. Nobody bought the post-mortem media twist rebuttal. People saw through the veil and no longer had to fear death, thanks to him. He changed everything and nobody could cover it up or change minds once the evidence was out there. So tell us, what do you think of the right to end one's life, doctor?" They almost never called me that.

I collected my thoughts, trying to remain in character. I replied, "I've seen a lot of death, I imagine more than most folks. When I was young the suffering shocked and appalled me, but then I became a professional, and somehow convinced myself that my medical training had immunized me from the suffering. But then when my wife became ill, the clinician within took a vacation. I was once again horrified at what we go through when disease eats away at our bodies, and at the suffering she went through.

"She begged me to help her die, but at that time my sense of honor to the Hippocratic oath somehow trumped my sense of honor to my bride. I watched her struggle through her last breath and expire, and I had the power to have given her the mercy she asked for and deserved. It was all such a waste, such a sorrow to experience. She deserved better, she should have died with dignity. Now my most profound memories of her are from that time, the last weeks of her life. None of the memories of the wonderful life we shared comes up without that monster shoving them down and eating them up.

"That's why I gave it all up. The injustice of what we are forced to endure at the end of life needs to be reconsidered. I couldn't be a part of a profession that says it is dedicated to the health of the patient and then denies them a choice in whether they want to continue care or not.

At that moment I realized I had said too much. The couple looked at each other. I thought I detected a knowing smile between them. I felt that this smart couple could have delved deeply into the subject of my research and its author. They could have, back in the day before the great deletions, done an internet search and read my biography, looked at my pictures, and found out about prominent events in my life such as the passing of my wife and more. It wouldn't have been difficult to figure out who I really was, if my being dead hadn't been so newsworthy of late.

But they didn't betray my secret, nor reveal their suspicions to me either. The subject of suicide and their possible guess as to my identity both died there that night.

25.

As the sailboat came north from the Panama Canal to cruise the shores of Pacific Mexico, I felt the urge to shift gears and revisit the places where my wife and I had shared our peak moments before her diagnosis. I told Zack and Nikki that I'd be getting off the boat in Puerto Vallarta and going inland from there. Our parting was especially brutal for me, as even though we all promised to keep in touch and visit, I knew that I would never connect with them ever again. The contact information I shared was made up, and as I stepped off the boat that balmy evening, a harder version of me walked away without looking back.

I didn't get a hotel or stay around to sightsee. A 20-minute walk brought me to the local bus station, the one where the peasants travel with their chickens in ancient school buses that meander up the coast to a symphony of squeaks and thuds from the potholes while they make clouds of black smoke as they strain to reach each small summit. What would have taken seven hours by a regular coach took almost a full day with the stops at almost every crossroads as passengers came and went. The ride rebooted my being and shook me into a state of vigilance of my environment.

In Mazatlán, I bought a ticket for the ferry that crosses the sea of Cortez to Pichilingue, a small port on the outskirts of La Paz. I resisted the urge to stop at an internet café and check for news about my whereabouts. The ferry is primarily a truck carrier, providing a shortcut for freight bound from mainland Mexico to the desert peninsula of Baja California. Passengers are an insignificant percentage of the fares collected on these daily crossings. I was unnoticed as I boarded, the ticket issuer and later the collector didn't so much as look directly at me as I transited the terminal and onto the ship. No ID was required for passage, giving me a comforting cloak of invisibility as I traveled.

The ride took 18 hours, including a two-hour stop in Topolobombo, up the mainland coast. In that port I was able to hop off the boat and walk to a pharmacy that had a computer for use. I sat down for a moment and wrote to Max, letting him know I was well and rested, and ready to resume the battle in a manner as yet unknown. After 15 minutes without a reply I returned to the ship in anticipation of my arrival in La Paz.

As I rolled with the motion of ship, I thought about the possibility of staying in La Paz until I had decided on a course of action, as long as I felt secure there of course. Being there would surely reconnect me to the wonderful memories I had of the time Olivia and I had shared in Baja. It would be therapeutic for me to resurrect and embrace them, now that I knew she was at peace. The love we shared would be my new primary recollection of her and us, the guilt and pain would no longer dominate my thoughts. She was free of pain and suffering and had forgiven me for my part in it. And because of the discovery, I must surely have given that same sense of comfort to many thousands of the loved ones left behind when their beloved passed on.

I decided that in order to keep La Paz as a completely safe place, I would return to Topolobambo whenever I wanted to communicate or do any other business such as receive funds. The following morning as the ship slowed as it approached Pichilingue, I was greeted with the distinctive beauty which defines the western coast of Baja California. The sea became an enchanting shade of tiffany blue as it met the brilliant white sands of the shore, and just beyond the shore watching over the scene, the desert hills and cliffs dotted with cactus, bizarre trees and brush, red and tan boulders, and little else. The contrast of the harshness of the terrain and the welcoming beauty of the shore makes this place unique in my travels. The sense of isolation was palpable and most welcome.

The Baja Peninsula has one two-lane road running its length, the 1,100-mile Trans-Peninsular highway, which was only completely paved recently. Parts are inhospitable, suggesting the prudence of four-wheel drive vehicles. In the monsoon season the roads are routinely washed out, taking days or weeks to become comfortably passable again. Because of this, and the limited economic opportunities therein, Baja California has remained one of the least populated places on earth, with an average of 27 people per square kilometer. And if you were to discount the 2.4 million people living in the border city of Tijuana, the average population per square kilometer drops to just 12.

There, computers are just now catching on. Hospitals use school composition notebooks to set appointments, and their information center there is a large room filled with patient file folders. Schools and government offices operate the same way, so getting things done takes much more time. The police are sociable and happy to have a chat and accept a small tip in lieu of issuing citations. For my purposes, however, this retro version of life would provide a comforting level of cover for the fugitive I had become.

There is a community of foriegners living in southern Baja, mostly retired Americans and Canadians living in Cabo San Lucas to the south and spreading out to the town of Todos Santos on the West coast and La Paz on the East. One more foreigner going about his business there wouldn't draw anyone's attention. I felt I should be safe there for a while.

Once in La Paz I found lodging in a quiet hotel a few blocks from the heart of town and the waterfront promenade locally known as the Malecon. From there I explored the town and discovered few changes since my last visit with Olivia. It was still a sleepy, tranquil place where growth comes at a carefully measured pace. I reacquainted myself with this lovely place and eventually found an apartment for rent a couple of blocks from the Malecon in a neighborhood sparsly dotted with foriegners, a small cluster of coffee shops, funky organic restaurants, a farmer's market, A European style bakery filled with rustic breads and sweet delicasies, and most everything else I would need to live, within a four-block radius. I soon settled in to my new invisible life.

26.

It had been more than a year since the death of my doppelganger in London. I felt a little less pursued every day-though I wouldn't let my guard down nor get too close to anyone here in my new home. There was a great sobering that came with those powerful people and institutions threatening me, then attempting to hunt and kill me twice, possibly killing my associates, and removing most of my scientific research and documentation from public access and twisting that truth into something frightening to suit their desires to control the beliefs of the people of the world. Those people and institutions are without morality or conscience in the pursuit of their agendas. Society, truth, and the public interest are all secondary to their private goals.

I was encouraged by the sentiments shared by my sailing companions, that the people were generally able to discern the truth from the cover up, the fake reality from the spin the media pushed onto the public. But I also realized that there would be many, many people who were persuaded by the fabricated reports and still held fear of what awaits us when we die. It just shouldn't be that way; Dying has now been shown to be a wonderful and welcomed experience that shouldn't be feared. Our passing from this experience to the next is completely natural, and if understood and embraced for that, surely mankind would lead a happier, more meaningful life in our time on earth, and not obsessed with guilt and fear of what may or may not happen to us when we die.

Exposing the conspiracy by once again reiterating the facts, along with the story of Van Kleef, his demand to falsify my findings, the killing of my stand-in, and all the rest, grew into an overwhelming moral imperative for me. Perhaps I would have been safer if I'd just decided to let the forces of deceit convince me to corrupt my findings, or not publish this book in answer to their actions that destroyed that truth, but I couldn't let it go. I might not be able to expose the individuals behind the lies, but showing that I am still alive and the truth is once again disseminated in its original form will expose the fact that there are dark powers trying to hide it. And hopefully help to remove the doubt they have cast on the reality of what lies beyond this mortal plane.

After a month in La Paz I reconnected with Max, travelling by ferry to Topolobombo on the mainland, just to have a bit of a buffer between where I live and where

I made contact with the outside world. Although I had trust in Max, there were several links in our chain of communication that could be hacked, so prudence was my guide.

When we were able to connect, he gave me a synopsis of what had happened over the past months. My work had been systematically deleted from the internet and almost everywhere else, the consensus was that I was dead, though there was still activity in the intelligence community regarding my whereabouts. Sales of the device had passed 25 million units, and the rise in suicides had seemed to have peaked and stabilized. Still, the majority of folks wanted to cling to life as they suffered their last breaths, but it appeared that perhaps 20% had decided to end their lives before they were in the final days of intense suffering.

He said I was still collecting royalties on the devices, so I needn't worry about money. Max had arranged it so that I now had several bank accounts and resources with which I could feel comfortable, but my paranoia kept me on edge any time I went to withdraw funds. Financial transactions could be traced.

As much as I tried to be at peace, to feel ordinary and invisible, the idea of elite international hitmen searching the globe for a sign of my whereabouts crept into the forefront of my mind all too often. And although I no longer feared death, the idea that I may still be hunted was to say the least unsettling.

I made peace with the distinct possibility of a violent and sudden death. I imagined, among other scenarios, that one day I would be coming out of the local farmer's market, groceries in hand, and hear my name called. The slight turn of my head, which I had tried to train myself to resist, would start the clock on my last day alive. At first I resigned myself to the idea that there was nothing to do but wait. But after a while the waiting, the looking over my shoulder, so to speak, became boring. Nobody recognized me, no voice calling out "Micheal" or "Doctor Brown" to see if I would react.

I realized that I had a gift and a passion that needed to be utilized, and that hiding (while tanning and fishing on a desert peninsula) while waiting to be discovered was a pathetic waste of all I had to offer the world.

Deciding to find a new path where I could help people once again, I came up with a new plan that would utilize the skills and energy I had to offer. I asked Max to send me an identity with a medical degree attached to it so I could legally practice medicine again. Once it arrived I made preparations to leave Baja and start being proactive in helping relieve the suffering of others. But first it was time to set the record straight about me and my research and our discoveries, how the information and events got corrupted, and of my death/disappearance.

I approached Scott Locker after observing him for several weeks, and after he earned my trust, I gave him this story. I decided that once he had written it and I had reviewed it for accuracy, when it was on its way to the publisher, I would leave Baja and move on to where I could safely serve those in need.

There is a huge swath of the world not driven by capitalism, where there are no big brother CCTV and facial recognition systems “protecting” citizens, where the NSA isn’t scanning millions of those CCTV images with special secret software looking for what they have likely labeled as another terrorist suspect. This huge part of the world is flush with poverty and disease and suffering. There, Doctors and others with a genuine passion for helping their fellow man are welcomed without suspicion. Welcomed with open arms. There is a calling for me out there, not necessarily as an advocate for euthanasia, but more as a care giver and physician. I have a plan, and as much as I’d like to share it with you, giving out those details would surely cut that plan (and my life span) short. So, know that by the time you read this I will be elsewhere doing something important, in my own small way, for mankind.

And for you, dear reader, please know that there is nothing to fear in Dying. It is a glorious, liberating experience that you should embrace without fear as just another important step in your journey, in *our* journey.

As Ram Das said so eloquently, “Let’s walk each other home”.

AFTERWORD

After the final edits were made and Dr. Brown gave his approval to these volumes, we had a celebratory dinner. As he left my home, he shook my hand and said goodbye. His final words were given to comfort me, and to all of us: "I wish you a process of dying that doesn't overwhelm you, a moment of dying that you can be conscious of and embrace. I wish you a future incarnation in a plane of incredible light".

That was the last time I saw him. Although I barely knew him, I felt a genuineness and compassion that I've all too rarely experienced in others. I trust the truth and sincerity of his story, and believe he is out there somewhere, in a place where he can once again help people ease their suffering…

ABOUT THE AUTHOR

Better than Living is Scott locker's second book, having written a memoir entitled: "Living on the water", about his relationship with a wooden yacht converted into a restaurant in Mexico and later into a Boutique hotel in the San Francisco bay. Apart from making contributions to travel publications about the Baja Peninsula and consulting on living and doing business in Mexico, he is currently working on three other books.

He lives in La Paz full time, and can be contacted at scottlocker@icloud.com

Made in the USA
Monee, IL
13 September 2026

62979553R00138